Blockchain and Digital Twin Enabled IoT Networks

Privacy and Security Perspectives

Editors

Randhir Kumar

Department of Computer Science and Engineering
SRM University, Amaravati, AP, India

Prabhat Kumar

Department of Software Engineering
LUT School of Engineering, Lappeenranta, South Karelia, Finland

Sobin C.C.

Department of Computer Science and Engineering
SRM University, Amaravati, AP, India

CRC Press
Taylor & Francis Group
Boca Raton London New York

CRC Press is an imprint of the
Taylor & Francis Group, an **informa** business

A SCIENCE PUBLISHERS BOOK

First edition published 2025
by CRC Press
2385 NW Executive Center Drive, Suite 320, Boca Raton FL 33431

and by CRC Press
4 Park Square, Milton Park, Abingdon, Oxon, OX14 4RN

ISBN: 978-1-032-51751-3 (hbk)
ISBN: 978-1-032-51753-7 (pbk)
ISBN: 978-1-003-40379-1 (ebk)

DOI: 10.1201/9781003403791

Typeset in Times New Roman
by Prime Publishing Services

Preface

IIIIIIIIIIIIIIIIIIIIIIIIIIIIIIIIIIIIIIIIIIIIIIIIIIIIIIIIIIIIIIIIIIIIIIIIIIIIIIIIIIIIIIIIIIIIIIIIIIIIIIIIIIIIIIIIIIIIIIIIIIIIIIIIIIIIIIIIIIIIIIIIIIIIIIIIII

In recent years, the convergence of blockchain technology and Digital Twin (DT) applications within the Internet of Things (IoT) ecosystem has sparked a transformative wave in how we perceive, manage, and secure interconnected systems. This preface sets the stage for a comprehensive exploration of the privacy and security perspectives inherent in the amalgamation of blockchain and Digital Twin technologies within IoT networks. The Internet of Things, characterized by the seamless interconnectivity of devices, sensors, and actuators, has ushered in an era of unprecedented data generation and exchange. However, this proliferation of data has brought about significant concerns regarding privacy and security. As IoT devices continue to permeate various facets of our daily lives, from smart homes to industrial processes, safeguarding sensitive information becomes paramount. Digital Twins, virtual replicas of physical entities or systems, have emerged as a pivotal component in enhancing the efficiency and effectiveness of IoT networks. By providing real-time insights, predictive analytics, and the ability to simulate and optimize operations, Digital Twins contribute to informed decision-making and proactive management of IoT ecosystems. The integration of blockchain technology into IoT networks introduces a decentralized and tamper-resistant ledger that ensures transparency, immutability, and trust in data transactions. Blockchain's distributed nature mitigates single points of failure and unauthorized access, addressing some of the security concerns associated with centralized systems. Moreover, smart contracts within blockchain networks automate and enforce predefined rules, adding an extra layer of trust and accountability to transactions. However, this amalgamation of technologies also raises intricate challenges, particularly in the realms of privacy and security. Balancing the need for transparency with the imperative to protect sensitive information becomes a delicate task. As Digital Twins generate intricate replicas of physical entities, ensuring the privacy of the data they handle becomes crucial. Simultaneously, securing the decentralized infrastructure of blockchain networks from potential vulnerabilities and attacks is imperative for maintaining the integrity of the entire system. This preface sets the tone for a deeper exploration of the multifaceted dimensions surrounding the intersection of blockchain, Digital Twins, and IoT networks. Through a nuanced examination of privacy and security considerations, this compilation aims to provide insights, solutions, and a holistic understanding of the evolving landscape where these technologies converge. As we delve into the subsequent chapters, we invite readers to embark on a journey that navigates the intricate terrain of safeguarding privacy and fortifying security in the era of Blockchain and Digital Twin Enabled IoT Networks.

Contents

CHAPTER 1

BCECBN

Blockchain-enabled P2P Secure File Sharing System Over Cloudlet Networks

Pabba Sumanth, Sriramulu Bojjagani,* Popuri Poojitha,
Ponnam Bharani, Thokala Gopal Krishna and
Neeraj Kumar Sharma

1. Introduction

Since the development of the internet, education has been much more accessible to everyone. People from many walks of life, including students, academics, and working professionals, use online communities to publish and sell their work. To participate in any of these services, however, membership is essential. The files are uploaded directly into the application; thus, there is no protection. Data Integrity verification is necessary to achieve security in cloud-based information systems [1]. Kumar et al. [2] developed a privacy-preserving and secure framework with the help of block-chain. In their approach, PPSF is working mainly on two key strategies: one is two-level privacy, and another one is an intrusion detection scheme. In 2017, Omar et al. [3] designed a privacy-preserving platform for healthcare data for supporting security features and assuring pseudonymity by using cryptography functions to protect the patient's data. There have been attempts. Users must subscribe to their application to see their work, and only those who have done so are

Cyber Security Lab, Department of Computer Science and Engineering, School of Engineering and
    Sciences (SEAS), SRM University-AP, Amaravati, Andhra Pradesh, 522240, India.
* Corresponding author: sriramulubojjagani@gmail.com

granted access to the associated files. Since third-party applications are used in this method, the content creator receives less payment.

To do away with intermediaries and ensure safe file sharing, we developed a revolutionary approach: a cloud-based system based on Block Chain. This method allows users to share files safely and swiftly [4].

The blockchain-powered cloud-based file-sharing system uses the cloud computing infrastructure to provide reliable and secure services. Cloud computing has completely changed the game's rules in the modern world. As a result of cloud computing, users and businesses may have access to software without investing in new computers [5]. Everything from making a computer-generated image to providing AI services might be done on the cloud. As a relatively new technological development, cloud computing has steadily gained market penetration over the last three years. With the cloud-based file-sharing method, users may easily share and receive data from one another [6]. This approach allows users to store and share data using cloud networks simply and efficiently. When we store information in the cloud, we give up control, which raises severe worries about the safety of our data's privacy and reliability [7]. This paper aims to go through an encrypted method of cloud-based file sharing. During this research, we developed a brand-new strategy for exchanging digital media. This file-viewing transaction is executed through a blockchain, which moves it from user to user. It allows many individuals to share and receive files without worrying about security. It is recommended that data be encrypted before cloud storage. This is possible due to the usage of an encryption technique. Most file-sharing services and applications implement secure file-sharing by ensuring only authorized users can see and download data.

Video lessons, research papers, and anything in between may all be uploaded in this manner by users. This encryption method works with every file type imaginable. Our primary focus is removing barriers between you and the original originator of a file. To do this, blockchain technology is utilized. The term "blockchain" refers to a distributed digital ledger maintained by a computer network that records and verifies financial transactions. In addition, the network is secure against attacks, and the information entered cannot be modified. A user who needs access to these resources can pay for them by submitting a transaction on the blockchain. The technique involves the creator of the file setting the price. Young individuals who do not know each other engage in this conversation.

On the other hand, features such as confidentiality, integrity, availability, and authentication must be ensured to safeguard shared information's privacy, authenticity, and truthfulness [8]. Bojjagani et al. [9, 10, 11, 12] addressed the various security attacks on three application levels, communication, and device. They demonstrated how to keep the information secure from intruders in their approach, especially with the well-known active attacks of Man-in-the-Middle (MitM), replay, Denial of service (DoS), and phishing attacks. And provide vulnerabilities with risk behaviours in their scheme.

Our key contributions to this chapter are the following:

1. We have developed a unique blockchain-based preserving and securing data services for IoT to assure user security. In this system, tamper-proof records generated using pairing-based encryption can be integrated into blockchain transactions. It allows tamper-proof data to remain secure even if the underlying blockchain is compromised.

2. Using blockchain-based smart contracts, we develop safe payment mechanisms that allow users and uploaders to pay for associated services reliably and automatically.

3. We present a security analysis to prove that our system is safe from collusion, tampering, and man-in-the-middle attacks. The findings demonstrate that our technique is efficient and practical, with a modest calculation overhead.

Organization of the chapter: The rest of the chapter is organized as follows: An overview of the related work on cloud-based privacy-preserving techniques using blockchain in Section 2. Section 3 describes the proposed model. Section 4 provides results and discussions, and Section 5 presents concluding remarks and future work.

2. Related Work

Multiple file-sharing software options exist. This generation of apps prioritizes openness over privacy and productivity. We need a centralized app where all files are accessible by every user who registers or uses the app because these apps share data. Therefore, we have highlighted a few related software programs:

We used the AES encryption technique in our app. It was formerly possible to encrypt information using AES. Rayarapu et al. [13] paid particular attention to the processes involved in secretly encrypting and storing data on a disc and then decrypting that data with the same key. This system employs a cryptographic algorithm based on the Advanced Encryption Standard (AES). AES-128, AES-192, and AES-256 refer to the relative key sizes (128 bits, 192 bits, and 256 bits) and rounds (10, 12, and 14) required to unlock the vault that closes around the data. The key and the encrypted data are transmitted via a series of characters that are exchanged in this technique. To prevent unauthorized changes to the encrypted files, they are made read-only. The system's main selling point is preventing users from deleting encrypted data using the right-click menu. As a result, there will be a higher level of protection for the data stored on the disc.

Amin et al. [14] developed a framework for helping both user cooperation and resource sharing was given extensive treatment in this essay. They advocated for a P2P network with centralized connection management performed by a directory server. The directory server was used to help create the overlay network and guarantee a quick and effective search. They devised a plan to build and manage the overlay network that makes the most available resources while keeping latency minimal. For the last step, we use simulation to test how well the proposed framework runs. Oliveira et al. [15] developed a framework for the security options provided by a few cloud storage services, allowing users to share files without intermediaries and pay only for the space they use. They used Amazon Simple Storage Service (S3), Google Cloud Storage, HP Public Cloud, RackSpace Cloud Files, Windows Azure Storage, and Luna Cloud. The permissions of the services, their semantics, and the various access-granting mechanisms used to provide these permissions to users were all discussed.

Furthermore, a collection of techniques for safely moving information among various public cloud storage services was outlined. Several protocols were developed to facilitate communication between users of different cloud services by augmenting a predefined set of characteristics. We took this as a guide and saved our files with Azure Storage Services. By utilizing Azure's services, we can rest easy knowing that our data is safe. We used the Paturi et al. [16] paradigm to combine a blockchain with a user's deposit of a negligible quantity of crypto to the file's author in a single transaction. Smart contracts are used for this purpose. Researchers used blockchain technology to devise a mechanism for financially rewarding participants in a waste management system. They proposed a new AI-powered management system built on blockchain and the IoT. It uses smart contracts and blockchain technologies to reward people for putting trash in smart bins. Yeh et al. [17] designed a framework on a revocable and monitorable peer-to-peer file-sharing mechanism over a consortium blockchain. For their approach, they ensured file revocation in the decentralized environment. The data integrity of the executable peer-to-peer file-sharing tool is verified and generate a unique file authentication code for each "InterPlanetary File System (IPFS)" node is used to determine whether the file-sharing system is synchronized correctly. Also, Yeh et al. scheme integrates the autonomous smart contracts and Intel SGX hardware to obtain the monitorable merit.

3. Proposed Model

We introduced a secure file-sharing solution to protect cloud-stored files and data. Users of the proposed system can simultaneously exchange files with several other users. This software provides the highest possible level of safety for its users. Before uploading a file to the server, the user can encrypt it, as shown in Fig. 1.

When a file is posted to this application, all registered users have access to it and its metadata, including the uploader's name, the file's size, and the date it was uploaded. A user can only view an uploaded file by requesting it directly from whoever originally posted it. When users attempt to download content, they are sent to a crypto payment gateway. An email with the secret key is sent to the user to confirm a successful transfer of funds. After entering the secret key, the user will have access to the files and be able to view or download them. We developed the AES algorithm to encrypt data in our proposed system. A private key is generated while the data is being encrypted. Any file type (.docx,.pdf,.mp4) can be encrypted with this framework, provided the secret key satisfies essential creation conditions such as required length and complexity. A file encrypted with a specific algorithm will have its name changed to filename.txt, suggesting that it may be read as plaintext or deciphered using a text editor. Data in an encrypted file is safe against tampering. The proposed method's steps are shown in the flowchart of the encryption and decryption process in Fig. 2.

We built a safe online platform to exchange information. Users must sign up for the service before they can use it. Upon completing the registration process, the user is automatically sent to the login page and granted access. The whole functionality of the framework is now available to the user. It includes viewing, uploading, encrypting, and decrypting data. For security reasons, before a user may send another user an encrypted file, the file must be encrypted with the encrypt function. User-generated private keys are used to protect information throughout the encryption process. Once the encryption is complete, the user can submit the encrypted file, and its corresponding metadata is shown in Fig. 3.

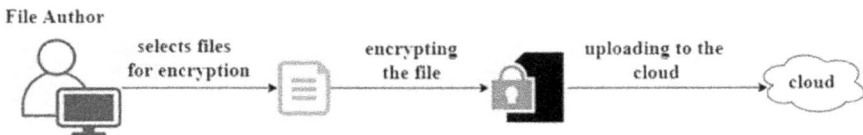

Figure 1. System architecture encrypting & uploading files to a cloud.

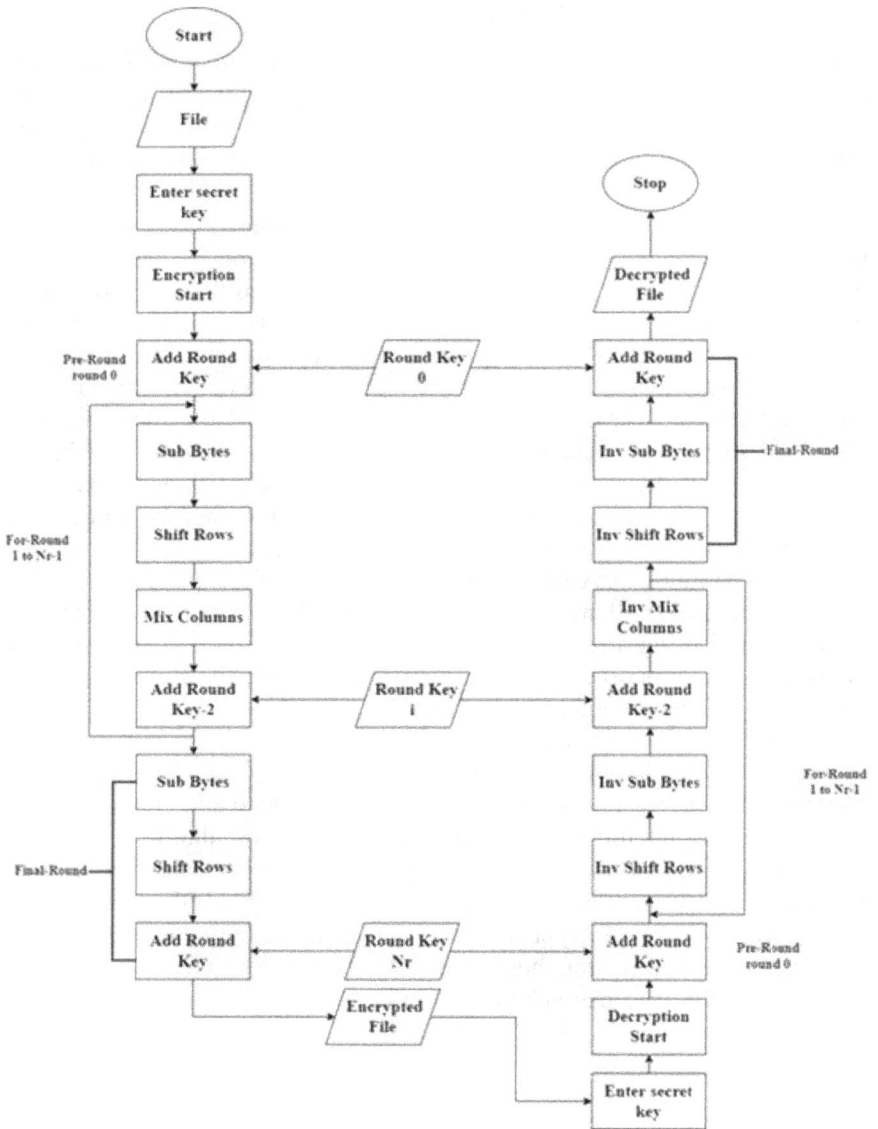

Figure 2. Flowchart of the encryption and decryption process.

Any user who wants to view a file after it has been posted will need the secret key produced by the person who submitted the file shown in Fig. 4. The user must use the same private key provided by the uploader to decrypt the file. If the key does not match, the user will be informed that the file cannot be decrypted and cannot access it. The user can view and download the file

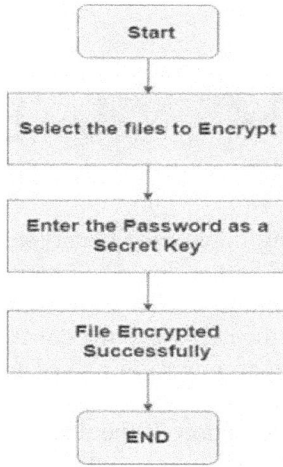

Figure 3. Encryption function.

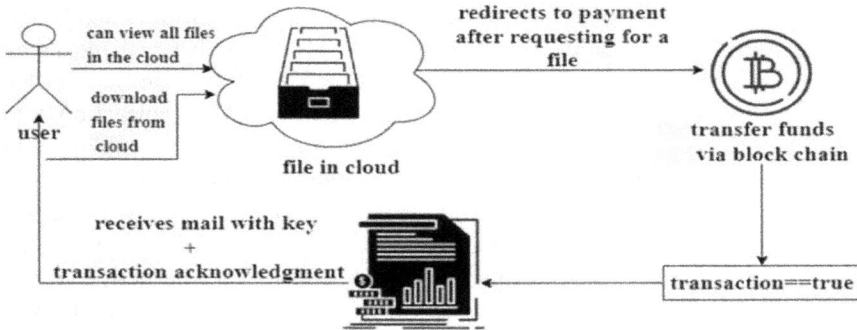

Figure 4. System architecture for retrieving the files.

successfully if the secret key is entered correctly. By utilizing this method, we can ensure the safety of the files stored on our server.

Any user who wants to view a file after it has been posted will need the secret key produced by the person who submitted the file. The user should provide the same secret key provided by the uploader to decrypt the file. If the key does not match, the user will be informed that the file cannot be decrypted and cannot access it. The user can view and download the file if the secret key is entered correctly. By utilizing this method, we can ensure the safety of the files stored on our server.

On view files, you may see all the uploaded documents. When you upload a file, it will be saved in the cloud. Once a user signs up for the service, they can access the application and decrypt any file shown in Fig. 5. User action

user

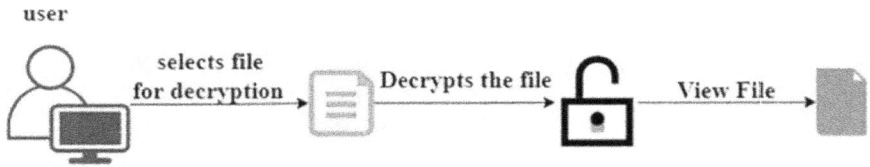

Figure 5. Flow chart for decrypting the file.

is required before accessing downloaded encrypted files. If the payment is successful, the user is sent to the download page after clicking the option to send money. A password-protected PDF with the key and download confirmation is emailed to the user when the download is complete.

The encrypted file is then manually picked for decryption. The data is sent to the decryption service for processing. We must enter the author's key to access the file during decryption. If the decryption is successful, the user can access the data. The decryption procedure is shown in the Flowchart in Fig. 6.

The financial transactions are conducted using blockchain technology. We use blockchain technology to keep track of all the file downloads and uploads that users request. The Pay to View feature is also realized using the blockchain bitcoin concept. The smart contract will execute these tasks. Blockchain is a distributed, immutable ledger that may record transactions and keep tabs on assets in a corporate or business setting. Rather than maintaining records in a central location, blockchain stores data in connected blocks shared across participants. Each block in the blockchain contains the cryptographic hash, timestamp, and transaction details from the previous block. Bitcoin's distributed ledger, or blockchain, was developed as an alternative to the traditional financial system, in which institutions play the role of trusted intermediaries in facilitating authorized transactions and preventing fraud. To

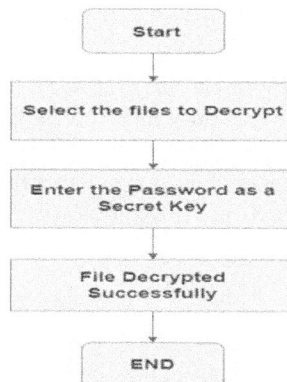

Figure 6. Decryption function.

maintain their customer database, banks impose astronomical transaction costs and maintain a personal database. Ether, Ripple, and Lite-coin are some of the numerous cryptocurrencies that may be used on the open-source blockchain network that facilitates transactions today. Payments sent to individuals may be guaranteed to arrive promptly and securely through Smart Contracts. Blockchain technology underpins the whole operation, ensuring openness in all dealings. Using a blockchain to transmit money helps make the software more transparent, letting the creator see who sent the money and received the data. All financial dealings are recorded on the blockchain for posterity.

Transactions involving money or other assets may now be conducted digitally and openly, thanks to smart contracts and agreements between two parties on a blockchain. Ethereum's smart contract functionality and decentralized blockchain architecture make it a frontrunner in the emerging smart contract market. The term "smart contract" refers to a small piece of code stored on a blockchain that runs automatically when specific conditions are met. This operation runs on its own. They streamline enforcing agreements, ensuring all parties know the conclusion without delays or extra effort. All financial dealings may be finalized using smart contracts, eliminating the need for any third parties. In the same way that a traditional contract sets down each party's obligations to an agreement, smart contracts do the same. Still, they also enact those obligations without needing a mediator or arbitrator.

A blockchain-powered cloud storage service keeps a log to track all the requests for downloading and uploading files. The Pay to View feature is also realized using the blockchain bitcoin concept, shown in Fig. 7. The smart contract will execute these tasks. If the uploader signs a Blockchain transaction, the application will allow him to upload his file. He will not be charged anything extra, but the user's confidential data of wallet address,

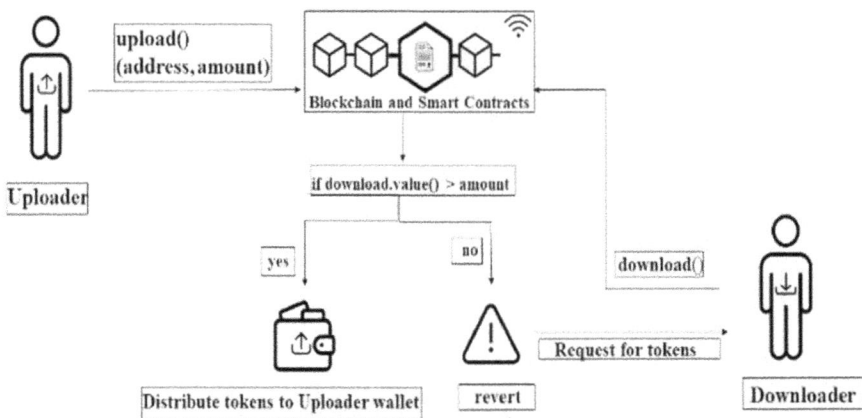

Figure 7. Flowchart of processing pay to view.

for example. Where he expects to get his crypto tokens from downloaders, royalty amount, data, and time will be recorded. The secret document is then uploaded to Azure DB, where anybody may download it.

Users can go through a directory of accessible files and download anything. They'll have to fork out the specified amount in bitcoins that the file's uploader stipulated. After a transaction is completed and money is received, the blockchain is updated with the downloader's information and the time and date the file was obtained. This creates a public blockchain log of file transfers. Payments are processed swiftly, and the identities of those who upload and download information can be reliably tracked, all thanks to blockchain technology.

As was mentioned before, we hosted our app on the Microsoft Azure cloud shown in Fig. 8. Microsoft Azure is a cloud computing service developed and managed by Microsoft. When it comes to cloud services, Azure has you covered. Like other Microsoft offerings, Azure is a free and open-source cloud computing platform; however, the Azure client software development kit (SDK) is not. Software as a service (SAAS), platform as a service (PaaS), and infrastructure as a service (IaaS), are all cloud computing paradigms that are provided by Azure. Numerous businesses take advantage of Azure's cloud services. While most of Azure's offerings are Microsoft-centric, some Linux-specific options exist. There is no need to install any software on the client machine to make use of Azure's hosting services for email servers, web servers, applications, Active Directory, virtual machines, remote desktop management, content delivery networks (CDNs), messaging services, data management, and big data tools, among others.[1]

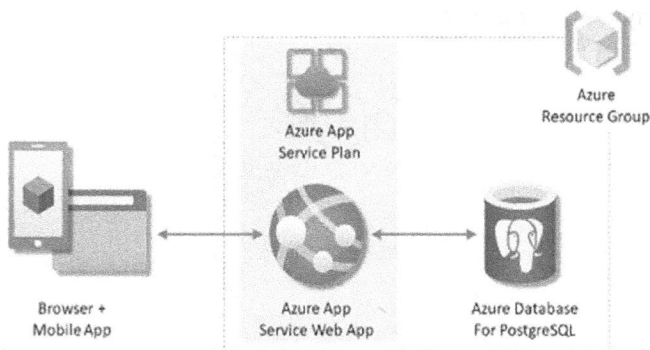

Figure 8. System architecture after deploying the application in Azure.

---

[1] Azure App Service. App service overview. https://learn.microsoft.com/en-us/azure/app-service/overviewapp-service-on-linux (Accessed on 6th Feb 2023).

We developed the framework on local machines, tested it thoroughly, and only then sent it to the cloud, where it is currently being hosted and used in production. However, deploying an app is difficult; it requires several behind-the-scenes processes to be finished. The Azure App Service, which we used to push our code to the cloud, requires Python 3.7 or later and a Linux server environment, which we couldn't utilize because we relied on the Django framework.

With Azure App Service, you may use any language or framework to build and host REST APIs, backend processes, and web apps. The Azure app service allows us to set scalability and availability parameters without worrying about managing the underlying infrastructure. In this chapter, we allowed an Azure App Service as a hosting provider since I am responsible for many tasks while operating my application locally, including:

1. We must specify and acquire actual servers, storage, networking equipment, and any other essential gear.
2. Ensure that the primary, backup, cooling systems, and so on are all operational.
3. Set up and configure the network.
4. Install and configure any virtualization software, operating system, middleware, or runtime components your application requires.
5. Install and configure a web server such as IIS, Apache, or Nginx.

As Azure is a PaaS, we can eliminate all these positions by moving app development to the cloud. Our team can solely focus on maintaining our app and its associated data. Azure takes care of the rest. You won't need to worry about managing the network or the underlying infrastructure. Examples include updating the OS, implementing critical repairs, updating runtime components, or putting middleware in place. Everything runs well because of Azure. Now you can devote even more time to the parts of the application that matter.

Before you can host our software in Azure, you'll need to create a web app for Azure App Service. If we have access to the Azure portal, we can use it to build an Azure app. All the required settings must be adjusted, including the resource group, web app name, application environment, app service plan, etc. If everything checks out, we can start building the web app. The next step is to create our web application. To move the database from on-premises to Azure, we must first make a PostgreSQL database in Azure after launching the web app. We added a firewall rule to our app's server after we created the Azure Database for PostgreSQL. Using Azure's web app and PostgreSQL, we establish a connection between the two. The web app code requires the value of four environment variables (DBHOST, DBNAME, DBUSER, and DBPASS) to connect to the PostgreSQL server.

After connecting to the PostgreSQL database, we push our application code to Azure. When an application is successfully deployed to Azure, a DNS name is automatically generated for remote access via the internet, allowing us to log in from any device.[2]

4. Results and Discussion

We utilized Microsoft's Azure App Service to deploy proposed framework to the cloud. After an app has been deployed to Azure, a public URL is created to be accessed from anywhere. We used the Remix platform and its associated web3.js API to interact with the Blockchain network during the deployment. We conducted our business dealings with the help of the meta mask plugin wallet and its associated test accounts. To set up our system, we resorted to Matic networks.

Figure 9 reveals that we must encrypt the file before uploading it to the cloud. That's why we experimented with a range of file sizes to gauge the time required for the encryption process.

The next step is to upload the encrypted data to the cloud. The size of the file will grow after encryption. After determining how long it would take to upload the encrypted files. The time in seconds vs. file upload is shown in Fig. 10.

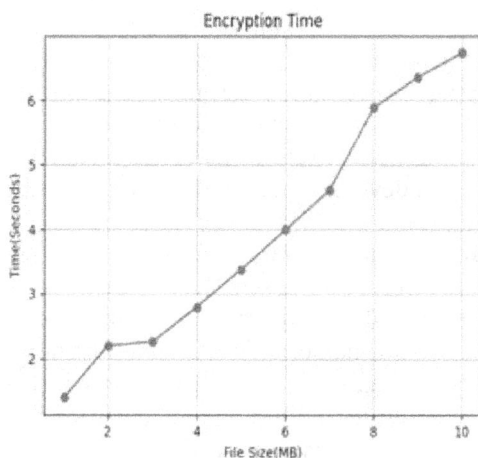

Figure 9. Encryption time.

---

[2] PostgreSQL in Azure. Postgresql in Azure. https://learn.microsoft.com/en-us/azure/app-service/tutorial-python-postgresql-app?tabs=flask (Accessed on 8th Feb 2023).

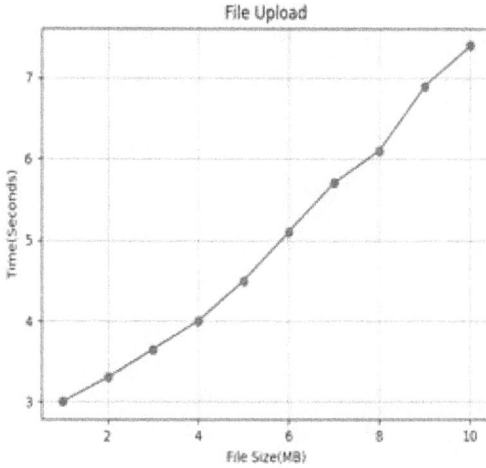

Figure 10. File upload.

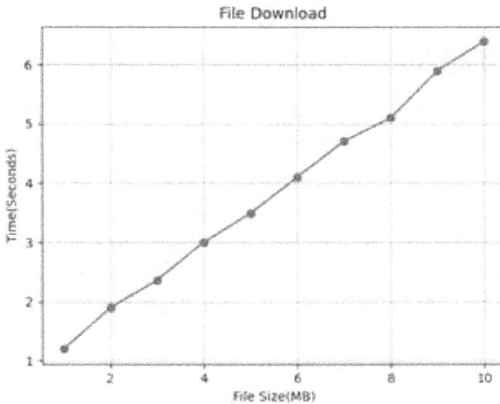

Figure 11. File download.

As seen in Fig. 4, after the information has been uploaded to the cloud, it is accessible to anybody who has registered for an account. As a result, the user must first send the crypto before the file can be downloaded. When the payment is complete, the user may access the download link. This has resulted in a wide range of modifications to file sizes. So, we timed the downloads to see how long it took. The file download vs. time in seconds is shown in Fig. 11. Since this is the case, decoding the information will follow data collection. We see the decryption method in action in Fig. 6.

Consequently, we tried out the function with various sizes to understand how much time is needed for decryption. Encryption results in a larger file

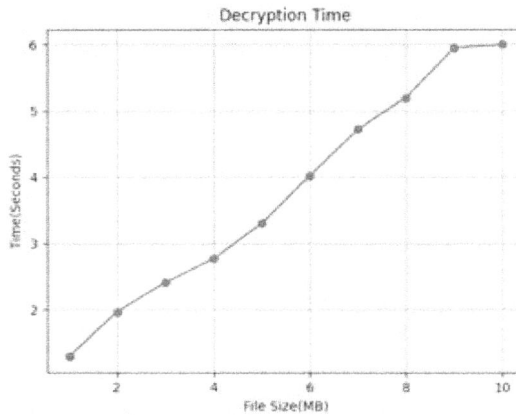

Figure 12. Decryption time.

size, while decryption is much quicker than encryption. This data was gleaned from the chart. The file size vs. decryption time is shown in Fig. 12.

5. Conclusions and Future Work

File sharing using cloud-based systems in peer-to-peer is challenging because many untrusted parties may access legitimate information. For this, many security and privacy-preserving schemes have evolved. But still, it is a task to save secure data from the adversary. In this chapter, we have proposed a novel solution so that no untrusted parties and attackers challenge to exploit the attacks during data share, especially in the cloud, with the help of a blockchain-enabled mechanism for file-sharing systems. Our main contribution is developing a secure framework to showcase novel architecture, secure mechanisms, and privacy-preserving using blockchain technology.

Before launching any application in the public storage areas, the client must establish a secure connection with a Content Delivery Network (CDN). It is essential to connect the millions of people who rely on file sharing; thus, we need a reliable solution to accommodate them all. For future work, the proposed framework is to integrate the developed application with a CDN.

References

[1] Zhao, Q., Chen, S., Liu, Z., Baker, T. and Zhang, Y. 2020. Blockchain-based privacy-preserving remote data integrity checking scheme for IoT information systems. Information Processing & Management 57(6): 102355.

[2] Kumar, P., Kumar, R., Srivastava, G., Gupta, G.P., Tripathi, R., Gadekallu, T.R. and Xiong, N.N. 2021. Ppsf: a privacy-preserving and secure framework using blockchain-based machine-learning for IoT-driven smart cities. IEEE Transactions on Network Science and Engineering 8(3): 2326–234.

[3] Al Omar, A., Rahman, M.S., Basu, A. and Kiyomoto, S. 2017. Medibchain: A blockchain based privacy preserving platform for healthcare data. pp. 534–543. In International Conference on Security, Privacy and Anonymity in Computation, Communication and Storage. Springer.

[4] Ferrag, M.A., Derdour, M., Mukherjee, M., Derhab, A., Maglaras, L. and Janicke, H. 2018. Blockchain technologies for the Internet of things: Research issues and challenges. IEEE Internet of Things Journal 6(2): 2188–2204.

[5] Deepa, N., Pham, Q.V., Nguyen, D.C., Bhattacharya, S., Prabadevi, B., Gadekallu, T.R., Maddikunta, P.K.R., Fang, F. and Pathirana, P.N. 2022. A survey on blockchain for big data: approaches, opportunities, and future directions. Future Generation Computer Systems.

[6] Wang, J., Li, M., He, Y., Li, H., Xiao, K. and Wang, C. 2018. A blockchain based privacy-preserving incentive mechanism in crowdsensing applications. IEEE Access 6: 17545–17556.

[7] Liang, X., Shetty, S., Tosh, D., Kamhoua, C., Kwiat, K. and Njilla, L. 2017. Provchain: A blockchain-based data provenance architecture in cloud environment with enhanced privacy and availability. pp. 468–477. In 2017 17th IEEE/ACM International Symposium on Cluster, Cloud and Grid Computing (CCGRID). IEEE.

[8] Zhao, Y., Liu, Y., Tian, A., Yu, Y. and Du, X. 2019. Blockchain based privacy-preserving software updates with proof-of-delivery for the internet of things. Journal of Parallel and Distributed Computing 132: 141–149.

[9] Bojjagani, S., Brabin, D.R.D. and Rao, P.V.V. 2020. Phishpreventer: a secure authentication protocol for prevention of phishing attacks in mobile environment with formal verification. Procedia Computer Science 171: 1110–1119.

[10] Bojjagani, S. and Sastry, V.N. 2016. STAMBA: Security testing for Android mobile banking apps. pp. 671–683. In Advances in Signal Processing and Intelligent Recognition Systems: Proceedings of Second International Symposium on Signal Processing and Intelligent Recognition Systems (SIRS-2015) December 16–19, 2015, Trivandrum, India. Springer.

[11] Bojjagani, S. and Sastry, V.N. 2017. VAPTAi: a threat model for vulnerability assessment and penetration testing of Android and iOS mobile banking apps. pp. 77–86. In 2017 IEEE 3rd International Conference on Collaboration and Internet Computing (CIC). IEEE.

[12] Bojjagani, S., Sastry, V.N., Chen, C.M., Kumari, S. and Khan, M.K. 2023. Systematic survey of mobile payments, protocols, and security infrastructure. Journal of Ambient Intelligence and Humanized Computing 14(1): 609–654.

[13] Rayarapu, A., Saxena, A., Krishna, N.V. and Mundhra, D. 2013. Securing files using AES algorithm. Int. J. Comput. Sci. Inf. Technol. 4(3): 433–435.

[14] Amin, H., Chahine, M.K. and Mazzini, G. 2012. P2P application for file sharing. pp. 1–4. In 2012 19th International Conference on Telecommunications (ICT). IEEE.

[15] Oliveira, T., Mendes, R. and Bessani, A. 2014. Sharing files using cloud storage services. pp. 13–25. In European Conference on Parallel Processing. Springer.

[16] Paturi, M., Puvvada, S., Ponnuru, B.S., Simhadri, M., Egala, B.S. and Pradhan, A.K. 2021. Smart solid waste management system using blockchain and IoT for smart cities. pp. 456–459. In 2021 IEEE International Symposium on Smart Electronic Systems (iSES) (Formerly iNiS). IEEE.

[17] Yeh, L.Y., Shen, C.Y., Huang, W.C., Hsu, W.H. and Wu, H.C. 2022. GDPR-aware revocable p2p file-sharing system over consortium blockchain. IEEE Systems Journal.

CHAPTER 2

Integrating IoT and SDN
System Architecture and Research Challenges

Chandroth Jisi,* Byeong hee-roh and *Jehad Ali*

1. Introduction

The Internet of Things (IoT) is a networking infrastructure that connects everyday physical devices, appliances, vehicles, and other things linked to the Internet and interconnects to each other for exchanging data [1]. The IoT devices can sense and collect physical environmental changes like temperature, humidity, light, smoke, etc. The collected data are then organized and analyzed to make decisions, to share the collected data with authorized services or applications, and to interact with humans and/or other devices. The IoT makes human life much easier by automatically interacting, remotely controlling, and monitoring devices to improve the quality of life and reduce human efforts as much as possible [2]. Smart homes, Smart agriculture, Transport systems, and Education are widely used IoT industries.

In the past decade, IoT has been a profound innovation that intensifies comforts and enhances the quality of human life. However, several challenges need to be addressed. One of the challenges of IoT is Standardization, as the different IoT devices and applications use a wide variety of communication technologies such as WiFi, LTE, Bluetooth, Zigbee, etc. [3]. A standardized architecture is crucial for hassle-free communications. The Second challenge

Department of AI Convergence Network, Ajou University, Suwon, 16499, South korea.
* Corresponding author: jisichandroth@ajou.ac.kr

is the Reliability and hardware problem. For any IoT-related applications, a physical device is a must. However, people prefer low-cost and highly reliable devices. Making such highly reliable devices at a low cost is very challenging. Another critical challenge is the security issues [4]. Competent health care and video surveillance system usage are increasing very fast. It is imperative to keep all personal details secure. Many data breaching situations are happening around the world. Many Machine Learning (ML) algorithms take part in IoT security purposes.

These are some of the crucial problems related to IoT. There are many other challenges, such as interconnectivity, integration, resource allocation, etc. [5]. Traditional networking components and infrastructure cannot deal with the application-specific requirements of IoT. The different applications require different Quality of Services (QoS). For example, the smart healthcare system needs reliable and low-delay connectivity, whereas the video streaming application tolerates delay. So, it is crucial to design the network infrastructure in such a way that each application gets its own QoS. However, it is a challenging task to allocate different kinds of resources and QoS for each application and connected device. Because every application and device are connected through available resources and protocols, it is challenging to provide designated QoS for each application with the same network infrastructure [6].

To address these problems and QoS-related challenges, the Software Defined Networking (SDN) is a promising solution. SDN is a technology that can centrally control the network using software applications to improve the network performance. SDN centralizes the network control by decoupling the forwarding process (Data plane) from the control process (Control plane) [7]. The control plane consists of one or more controllers, where all the intelligence is located. The control plane can monitor, analyze, and make decisions about the network. The SDN infrastructure is more software-dependent and less hardware-dependent. That is, it is easier to program different applications and related QoS remotely without the help of new hardware equipment. With the help of SDN, the IoT network infrastructure will be more scalable, standardized, and secured [8, 9].

2. IoT

IoT is the new paradigm that connects physical objects, collects, and analyzes data, makes decisions, and controls devices remotely. IoT enhances the quality of human life by automatically controlling physical devices and using novel applications. Physical objects, devices, or sensors are the essential elements of IoT. These devices are capable of sensing surrounding environmental changes and collecting those data. The data could be pressure information of a person or a video from a surveillance system. Recently a device consisting of multiple sensors can collect heterogeneous data in real-time. The collected data is transmitted to neighboring devices, a central unit, or a cloud using various communication technologies such as mobile networks, Wi-Fi, Bluetooth, RFID, etc. [10]. The collected data is then processed in the storage location. The edge devices are also capable of processing small quantities of data. The cloud is the finest solution for storing and processing large amounts of data. Different Machine Learning (ML) and statistical methods are used for analyzing these data. The processed information then is sent to the user or appropriate devices for further action.

Most industries, such as semiconductors, telecommunications, and software, invest in IoT technology. The Finances [11] estimate that the number of connected devices will likely be 25.44 billion. Almost every industry is using IoT applications in their business and products. The major IoT-dependent industries are Health care, Smart home, Transportation, Manufacturing, and Education. Figure 1 shows the heterogeneous IoT applications and different communication technologies.

The healthcare system is one of the leading industries that depend on IoT technology. Wearable IoT devices can monitor the patient's data 24/7 and warn health care professionals in case of unusual data. It can analyzes the patient's data and diagnose the illness. IoT technology is used for other complex tasks like robotic surgery. Internet of Vehicles (IoV) is another IoT application for intelligent transportation systems. In IoV, vehicles are connected and with the environment through the internet. The connected vehicles can collect data from other vehicles and the environment to compute the best traffic route,

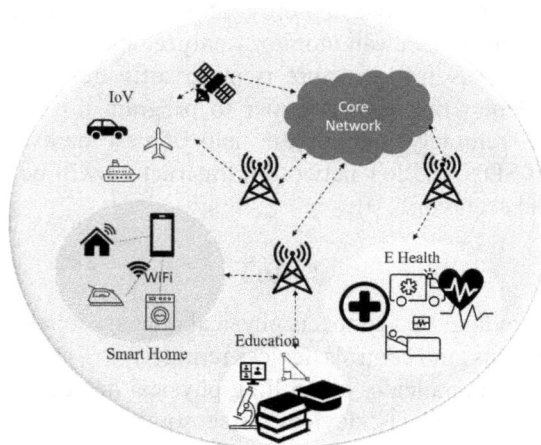

Figure 1. IoT applications.

avoid accidents and congestion, etc. Many other IoT-related applications that enhance the quality of human life [12].

2.1 IoT Architecture

Five-layer architecture is the best-proposed architecture for connecting heterogeneous IoT devices, communication technologies, and business aspects [1]. Five-layer architecture is the extension of the basic IoT architecture and is shown in Fig. 2. Next, we provide a brief explanation of the Five layers.

2.1.1 Data Layer

The data layer, or the perception layer, is the bottom layer in the layered architecture. The data layer consists of sensors and actuators that sense environmental changes and collect information like temperature, pressure, gas leakage, intrusion detection, etc. These physical devices are connected either through wired or wireless connectivity. The collected data is transferred to another layer for further action.

2.1.2 Network Layer

The network layer ensures connectivity between devices and the core network. It is the layer between the data layer and the data processing layer. The collected data can be transferred through various communication protocols such as RFID, Wi-Fi, Bluetooth, BLE, ZigBee, etc.

2.1.3 Data Processing Layer

The data processing layer, also called the middleware layer, can store, analyze, and make decisions based on the data. It stores the data with the device address, name, date, and sensed data. It manipulates and computes the data using ML/statistical technologies to find hidden information and make the appropriate decisions. Some of the responsibilities of the preprocessing layer are: (1) Data accumulation: Correctly assigning different data types to their appropriate storage space (2) Data abstraction: Aggregating the data from various sources (3) Data analysis: Finding the hidden pattern of data for decision making.

2.1.4 Application Layer

The application layer manages the different applications or services requested by the customers. For example, the application layer can check the whole house and inform the owner about the home's condition for security purposes. The Application layer provides high-quality services to the end user. It offers

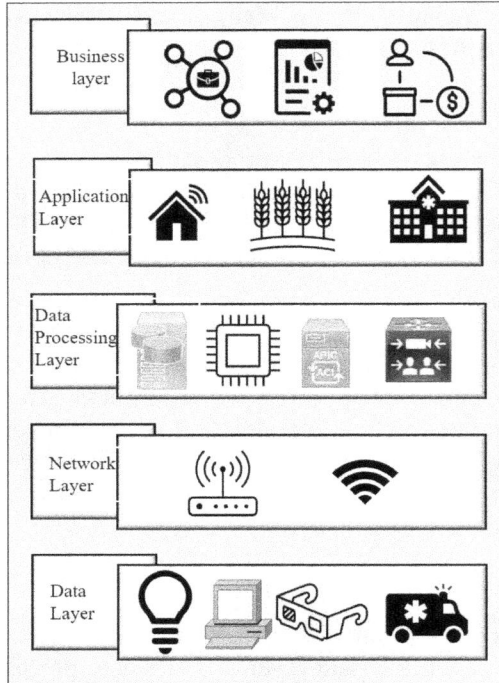

Figure 2. IoT five layer architecture.

various services, such as smart agriculture, intelligent healthcare systems, Transportation systems, and Education.

2.1.5 Business Layer

The business layer controls the application and services of the overall IoT system. The business layer does the business model of the services requested by a customer. Flowcharts and graphs can be used to show the performance of a business model. It also designs and implements the whole model, analyzes it, monitors, evaluates, checks the errors, and modifies the system.

3. Software Defined Networks (SDN)

Software Defined Networks (SDN), often referred to as a new paradigm, simplify network management and control by decoupling the control plane and data plane [13]. SDN enables the programmability of the network for automation. Computer networks include many physical devices (switches, routers, sensors, etc.) and protocols for the functionality of these devices. Device management and scalability are the main problems for traditional networks. The SDN-enabled network can easily manage the network by using logically centralized SDN controllers. The control logic from the networking devices is shifted to the SDN controllers, which enhances the flexibility and programmability of the network [14].

3.1 *SDN Architecture*

The SDN architecture [15] consists of three layers: the data layer, the controller layer, and the application layer, shown in Fig. 3. The SDN architecture also provides a set of Application Programming Interfaces (APIs), namely the Northbound and Southbound interfaces, for simplifying the network services.

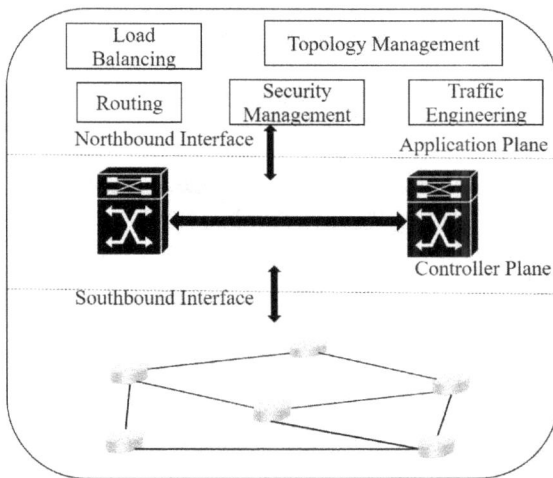

Figure 3. SDN architecture.

3.1.1 *Data Plane*

The Data Plane consists of network devices such as switches, routers, nodes, etc. The network devices are responsible for forwarding the data based on the rules implemented by the controller plane. The OpenFlow protocol is used for communication between the data plane and the controller plane. At the arrival of new packets, OpenFlow switches to check the 'matching fields' of the flow table entries and forward the data with the matched flow entry. If there is no matched flow entry, the switch sends a 'packet-in' message to the controller for further action.

3.1.2 Controller Plane

The SDN controller plane consists of logically distributed controllers, which provide intelligence to the entire network. The controller is responsible for forwarding policies and network management such as routing, load balancing, topology management, security, etc. The control plane communicates to the data and application planes through the southbound and northbound interfaces, respectively.

3.1.3 Application Plane

The application plane consists of different user-defined applications such as smart cities, innovative healthcare, transportation systems, etc. It communicates with the controller plane through the northbound interface.

4. SDIoT

Managing heterogeneous devices, communication protocols, and generated data types in IoT systems is challenging. Due to programmability, central view, and innovative protocols, SDN can be integrated with IoT systems for better performance and is named SDIoT [16]. SDN controller and OpenFlow switches are the main components of the SDIoT network [17]. OpenFlow switches are used to forward the data packets. SDN controller connects the switches and communicates through OpenFlow protocol. The SDIoT networks are used in health care, smart city, smart home, etc. applications to get full advantages of the central control from SDN [18]. SDN manages resource management, routing, and communication in IoT networks.

4.1 SDN-enabled Smartcity

The smart city enhances the life quality of people living in urban areas [19]. A large deployment of sensors and other physical devices connected with modern technologies gathered information from the environment and built intelligent functionality. The smart city relates to many other components, such as smart transportation, market, education, health care, etc., for their functionality. Security is the primary concern of innovative city applications. A secure smart city architecture is essential for protecting personal data, avoiding cyber-attacks, automation, and adding functionalities.

To enhance the scalability, security, and storage and reduce the energy consumption, latency, and collecting and organization of the data, the authors [20] propose a distributed and decentralized blockchain-software defined networking-based energy-aware architecture for smart city applications shown in Fig. 4. The perception layer consists of all IoT devices that provide the data to the users. The edge layer processes the data efficiently,

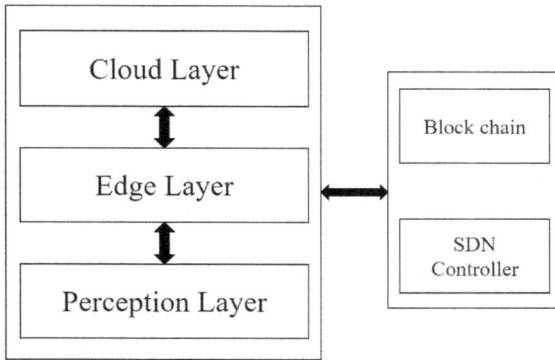

Figure 4. SDN enabled smart city architecture.

and the cloud layer stores the data and provides the service to the users. The centralized SDN controller can manage network activities and detect cyber-attacks, blockchain provides end-to-end security against threats, and network function virtualization (NFV) reduces the entire energy usage and increases the lifetime of devices. Moreover, a cluster head selection (CHS) algorithm cooperates with the network reducing energy consumption.

To improve the Quality of Service (QoS) of human life, protect personal data, and make intelligent decisions for various functions, the work done in [21] presents an SDN-based four basic IoT architectural blocks for secure smart cities. The blocks are the Black network, SDN controller, Unified registry, and Key management system. The Black networks can secure all data associated with the IoT protocols. Black networks avoid both active and passive attacks and ensure the security and privacy of the network. The SDN controllers ensure a secure communication path for black networks and other smart city applications. The Unified registry manages all the heterogeneous functions, addressing schemes, and physical devices. The Key management systems focus on the secure communication of IoT nodes utilizing shared keys. The keys enable security, simplicity, and resource efficiency.

4.2 SDN-enabled Smart Healthcare

The smart healthcare system enables the services to the patients inside the hospitals as well as outside of the hospital. The wireless body area network (WBAN) [22] is the key enabler for the above-mentioned service. WBAN consists of various sensors, such as blood pressure monitors, body temperature monitors, glucose monitors, etc., that can sense the health-related data of individuals and forward any unusual data to healthcare professionals. Delay and security threats in data communication are the primary concerns of smart healthcare systems. Any alteration in the patient's health data may have profound implications. A well-defined architecture is a must to overcome the issues in smart healthcare.

The Soft-Health [23], an SDN-based fog architecture, is proposed to serve various IoT-based healthcare applications. Soft-Health ensures minimum delay, less packet loss, and minimum network overhead. The fog layer consists of SDN switches that forward the incoming packets to the appropriate fog or cloud layers based on the criticality index (CI). CI is based on the physiological parameters of the sensed data. Also, Soft-Health utilizes the maximum capacity of the fog node through an optimization function.

To address the various healthcare applications and their QoS, the authors [24] propose a multi-tier healthcare architecture composed of end devices and edge and cloud servers. In this architecture, the resources are organized in a layered manner and controlled by SDN controllers shown in Fig. 5. The device tier consists of sensors and wearables that can sense medical data. The Edge tier makes the real-time analysis of sensed data, accesses the patient's past medical records, and forwards them to the health care professionals. It also stores these data temporarily. The cloud tier stores the data permanently for future use. The SDN controller layer controls and manages all these tiers through the northbound interface.

An intelligent software-defined fog architecture (i-Health) is presented in [25], which limits unnecessary data transmission in the Industrial Internet of Things (IIoT). The locally processed sensed data is transferred to fog or cloud for effective process and decision making, leading to unwanted delay. The i-health reduces this delay by placing fog nodes near the local data processing (LDP) unit. The proposed architecture finds the best accessible nodes based on their QoS, such as memory, battery capacity, I/O, and processing capability. Then, the Fog ranking Service (FRS) unit determines the rank of each fog node based on the QoS parameters.

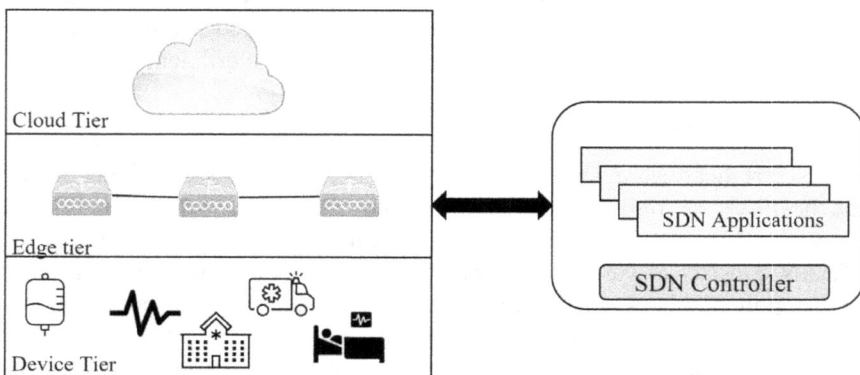

Figure 5. SDN enabled smart healthcare architecture.

E Barka et al. [26] present a Secured and Trusted Healthcare Monitoring Architecture Using SDN and Blockchain (STHM) to ensure a healthcare system's security, privacy, and flexibility. The STHM consists of four layers, namely: User layer, SDN layer, Security layer, and Blockchain layer. The user layer is the bottom layer, which includes all types of sensors, actuators, and mobile applications for sensing and communicating data to the target databases. The user layer is integrated with the SDN layer through southbound APIs and the Security layer through Northbound APIs. The security layer consists of various security modules which provide security and privacy for health monitoring systems. The Blockchain layer ensures security for all system actors.

4.3 SDN-enabled Smart Grid

The smart grid (SG), also called the smart electrical grid, is a modern power grid [27]. The system uses two-way digital communication for power generation. It efficiently generates and transmits power using machine learning (ML) techniques and other technologies. The smart grid provides various benefits to the users, such as self-monitoring, self-healing, and control.

SG and Electrical vehicles (EV) are increasingly popular due to their simplicity and energy availability. The EV market is increasing drastically. However, this will make an impact on the SGs. To normalize the power demand of SG, the work in [28] proposes a decentralized cloud computing architecture based on SDN and NFV. The users can communicate with the grids in a real-time way. The proposed architecture can optimize the energy, load, and price and maintain the grid's stability.

The SDN-microSENSE [29] shown in Fig. 6, is an SDN-enabled smart grid architecture for security concerns. The SDN-microSENSE consists of three main modules: (1) SDN-microSENSE Risk Assessment Framework (S-RAF) (2) Cross-Layer Energy Prevention and Detection System (XL-EPDS) (3) SDN-enabled Self-healing Framework (SDN-SELF) is deployed in the application plane of the SDIoT architecture. Apart from the application layer, the SDN-microSENSE architecture also includes (a) a Data Plane, (b) a Control Plane, and (c) a Management Plane. The Grid plane consists of various devices used in smart grid applications for collecting data. The data plane contains Openflow switches for forwarding data packets collected from the grid plane. The control plane consists of SDN controllers that get information from both the application and management planes for configuring the data plane accordingly. The application plane consists of different business problems related to industries. Finally, the management plane provides all the functionalities related to security.

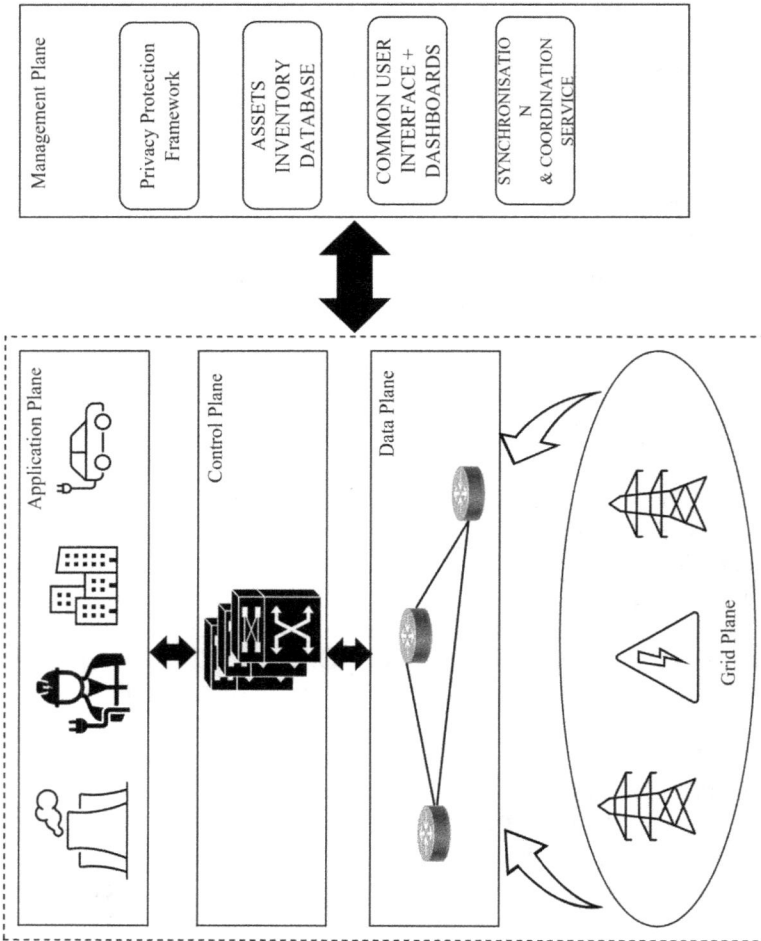

Figure 6. SDN enabled smart grid architecture.

4.4 SDN-enabled Smart Transportation

Smart transportation or Intelligent transportation system uses various technologies to monitor, analyze, and manage transportation system to increase efficiency and safety. Emerging technologies such as Fifth Generation (5G) communication and IoT make intelligent transportation a reality. The IoT provides sensors and actuators for collecting data, and 5G technology provides high-speed data communication. The main concern of intelligent traffic transportation development is avoiding accidents and traffic congestion in an energy-efficient and cost-effective manner.

To handle the traffic congestion problem in VANET, an SD-VANET [30] framework is proposed. The data plane consists of vehicles, sensors, and OpenFlow switches. The SDN controller is responsible for taking all the actions, such as routing, load balancing, topology management, congestion control, accident prediction, etc., based on the data from the data plane. For congestion control, SD-VANET periodically collects data from the data plane and builds a global network to make decisions such as route planning, mobility management, and congestion mitigation.

The work done in [31] provides a generic architecture of SDVN along with its different planes and their interconnection shown in Fig. 7. The Data Plane consists of SDN-enabled switches. The data plane is connected to the control plane through the southbound interface and forwards the data packets according to the controller's instructions. The control plane consists of an SDN controller, which is the brain of the network. The controller generates the flow rules based on the data packets. The application plane is the topmost layer connected to the control plane through the northbound interface. The

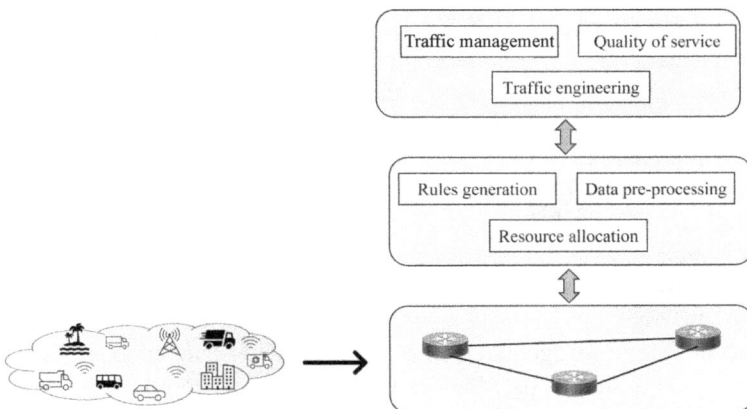

Figure 7. SDN enabled smart transportation architecture.

application layer consists of all the applications, such as routing, QoS, traffic management, etc.

4.5 SDN-enabled Smart Agriculture

The largest food source of the world is Agriculture. Based on the United Nations (UN) survey, the world population will reach 11 billion by 2050, meaning that the global demand for food will continue to increase [32]. Smart agriculture depends on technologies such as IoT, robotics, and Artificial intelligence (AI) for farming to increase the quality and quantity of the corps with less human intervention. The lifecycle of smart farming includes (1) observation, (2) diagnosis, (3) decision, and (4) action. In the observation phase, the sensors collect data from the crops, soil, or atmosphere. The collected data was then analyzed through various technologies. Based on the analysis result, the authority can decide, and the end user makes an action according to this, and the cycle repeats from the beginning. However, smart agriculture is still developing and faces many challenges, such as security, continuous monitoring of soil and water, automatic irrigation, etc. Compatibility with the technologies used in intelligent farming compatibility, constrained resources, and big data are also other challenges.

Various technologies such as cloud computing, SDN, and AI enable varied solutions to overcome these challenges. The SDN provides many advantages to smart agriculture, including network scalability, resource utilization, and application customization.

The authors [33] discussed a generalized SDN-based architecture for agriculture applications shown in Fig. 8. The architecture consists of Four layers. The bottom layer is the perception layer, which includes various IoT-based agricultural networks. The data plane is comprised of SDN switches dedicated to data forwarding. In the control plane, the SDN controller with the NFV orchestration system forms a centralized control system for the whole network. The application layer consists of different user-defined applications for agricultural services.

The work done in [34] focuses on the privacy and safety issues of Internet of Things (IoT)-based Precision Agriculture (PA). The proposed SDN gateway regulatory system ensures control of a foreign device without having access to sensitive farm information. The proposed framework is divided into two different devices and network layers. The devices are separated as home and foreign agents. Home devices are registered devices in the network. Foreign agents are devices that are not registered in the network. Foreign agents are controlled and managed by SDN controllers. SDN ensures complete security to the underlined network by blocking the threatful foreign agents.

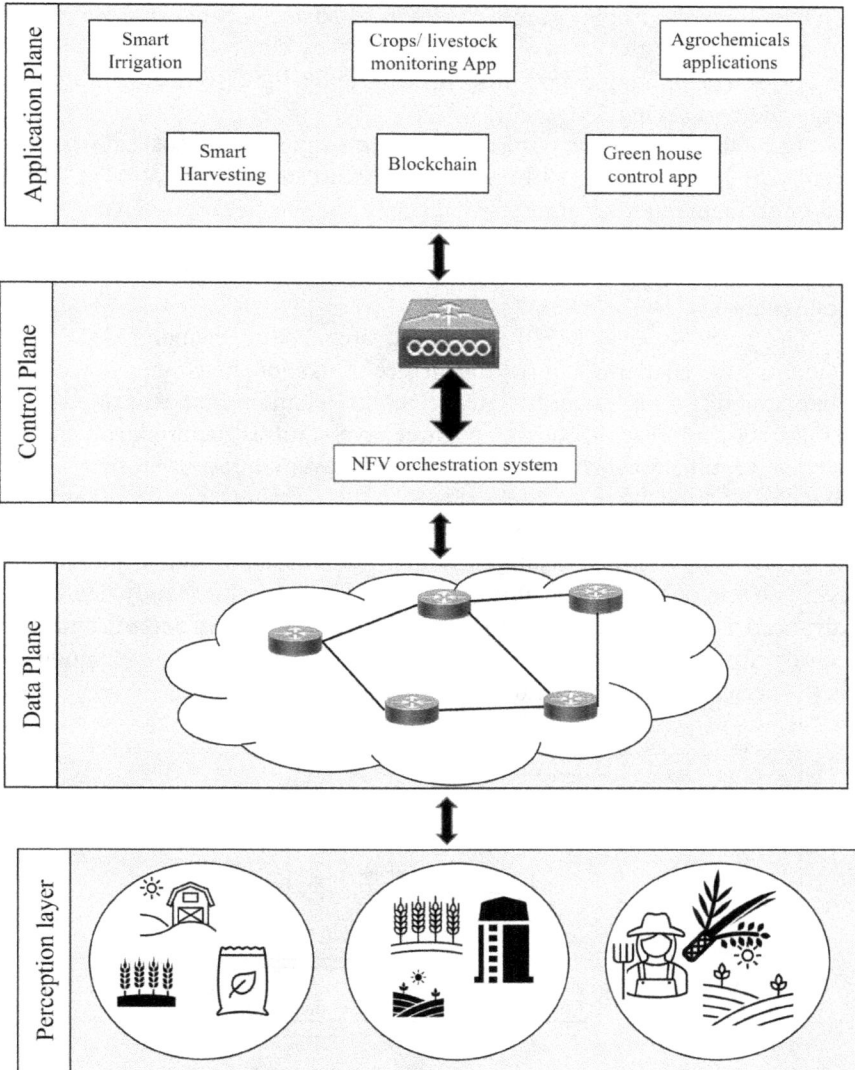

Figure 8. SDN enabled smart agriculture architecture.

4.6 SDN-enabled Smarthome

A smart home refers to an automatic and secured home environment where all appliances and devices are connected to the owner and controlled remotely from anywhere with the help of the internet. Smart home devices are connected to each other and controlled by a central point such as a smartphone, laptop, or tablet. Innovative home security systems can detect any unusual movement in the home when the owners are away, and they can call the authorities or the owner. The lighting control systems control the lights, reducing electricity usage and cost savings.

While the smart home ensures security, energy savings, cost savings, and Quality of life, there are still challenges. Smart homes always need strong and continuous internet access, which may require hardware access points for the entire house. Security risk is the other concern. Hackers can control the cameras and routers, steal personal videos, and ruin home appliances' functionality.

The work done in [35] proposes an architecture, namely SHSec, for addressing the challenges of flexible device utilization, heterogeneous device interoperability, and security enhancement in smart homes using SDN. SHSec, shown in Fig. 9, consists of three layers, infrastructure layer, control layer, and application layer. The infrastructure layer comprises all smart home devices and appliances, such as fridges, lights, washing machines, and gas sensors. The infrastructure layer is connected to the control layer. The control layer enables all the functionality and security of the smart home. The control layer consists of the orchestrator and KNOT. Orchestrator mainly deals with Advanced Persistent Threats (APTs) and provides access control and data security for application layer services. KNOT consists of three modules: packet parser, flow graph builder, and attack mitigation. It deals with security

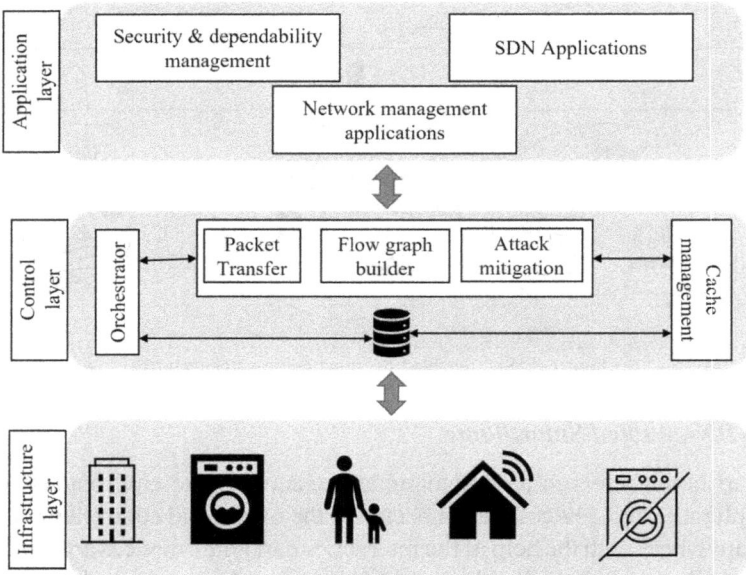

Figure 9. SDN enabled smart home architecture.

attacks and mitigates them when a saturation attack occurs. With the help of an orchestrator and KNOT, the SHSec provides flexibility and security to the home environment.

SDN with ML and Deep learning (DL) based architecture for smart homes is proposed in [36]. The architecture consists of three layers—the first is the data layer, which includes all the sensors and home appliances. The Raspberry Pi is used as a gateway for connecting all the devices. The middle layer is the OpenFlow layer, consisting of OpenFlow switches for data transmission. The control layer consists of a Ryu controller with Raspberry Pi installation. The ML or DL algorithm in the control layer classifies whether the incoming data packet is normal or abnormal. If the data packet is normal, the controller adds a flow rule for the forwarding process. Otherwise, for abnormal traffic, the controller adds a flag to the data packet and asks the switch to discard the packet. The controller can ensure the security of the entire home network by using ML or DL implementation.

5. Other Integrated Technologies in SDIoT

A high computation and communication capacity and extensive resources are required to meet the demands of future massive IoT applications. The SDN with emerging cloud-related technologies (fog and edge computing) can play an essential role in supporting and implementing IoT applications [37]. Moreover, SDN can integrate with the recent advent technologies such as blockchain and NFV to play a vital part in IoT application management and security. Fog and edge computing extends the cloud services to the edge of the network to reduce the latency of IoT applications. The NFV technology provides more resources to the IoT application through virtualization. The blockchain ensures the security and confidentiality of each data communication in the IoT network. Cloud computing provides ample storage space for massive IoT data. Moreover, the SDN technology integrated with IoT can manage all the technologies centrally.

5.1 NFV Integrated with SDIoT Networks

Network Function Virtualization (NFV) replaces the hardware equipment with Virtual Machines (VM). Due to the NFV, the network operators can provide new services and applications without installing new hardware components. So, they can enable the services with less time and cost. The virtual machines run all the software dedicated to routing, load balancing, etc., instead of the specific network device. The operators can program all the services in the SDN controller for VMs. The NFV technology reduces new application costs and speeds up the process. Security is the primary concern of NFV because it is still in its infant stage.

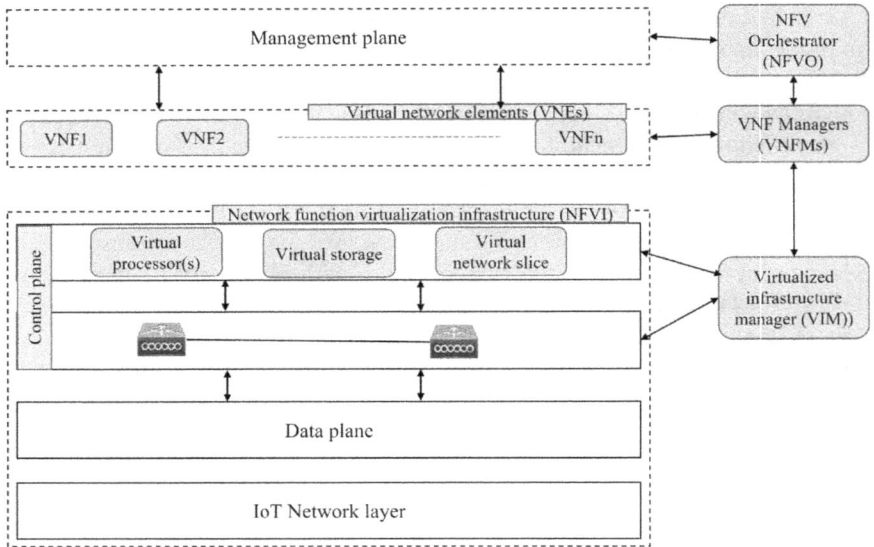

Figure 10. NFV integrated SDIoT architecture.

A general SDN-IoT architecture with NFV is shown in Fig. 10, is discussed in [38]. The architecture consists of different modules: (1) NFVI, (2) VNFs, and (3) Management and orchestration plane (MANO). NFVI includes different hardware and software components required to connect to the carrier network. VNFs are run on one or multiple VMs and provide functionality to the different applications and services. The MANO provides connectivity for different services to NFVI, VNF, and APIs.

5.2 Blockchain Integrated with SDIoT Networks

A blockchain is a database that is shared between all nodes in a network. Blockchain can store information in a digital format. Blockchain ensures the guarantee of data security in its database. The blockchain stored the data in a block instead of tables. The blockchain can collect all the information, but there is no facility to edit, alter, or delete. In this way, the blockchain ensures the security of the data.

IoT interconnects devices and services to collect data, discover information, and support automation in various fields such as healthcare, education, smart cities, and smart homes. IoT always deals with heterogeneous devices and services that also provide heterogeneous data. IoT must provide flexibility, strict security, and confidentiality for these applications. SDN offers the scalability and flexibility of the services. In parallel, the Blockchain

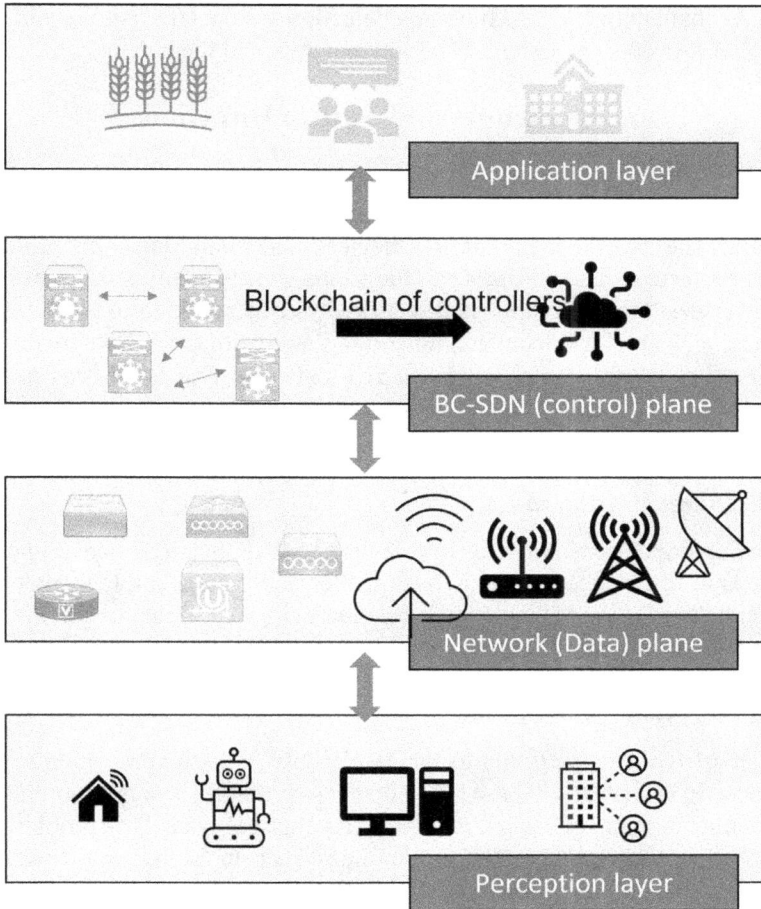

Figure 11. Blockchain integrated SDIoT architecture.

ensures the security of the system. The Blockchain-enabled SDIoT is now a new business model for various applications.

The work done in [39] proposes a blockchain-based architecture called BCSDNIoT for the SDIoT network for security, as shown in Fig. 11. The architecture consists of four layers. The perception layer includes all the IoT devices, sensors, and actuators for collecting data from different environments. The data plane consists of network gateways and OpenFlow switches. The gateway provides communication between IoT devices and OpenFlow switches. The control plane includes distributed SDN controllers and a blockchain network. All the controllers are connected to the blockchain network, so each device in the perception layer can also communicate with the blockchain network. The Intrusion detection system (IDS) module in each controller prevents any external and internal attack in the network.

6. Challenges and Future Directions

6.1 Security

Security and privacy issues are the main problems related to the SDIoT network. The heterogeneity of IoT devices and communication protocols worsened security issues [40, 41]. Due to the limited computing power and battery issues, highly complicated security systems added to IoT devices are impractical. The blockchain technique integrated with the SDN network can manage the security issues in the SDIoT network. The SDN provides the central view of the network, and the blockchain provides the data integrity in the network.

6.2 Resource Allocation

Allocating enough resources to IoT devices and applications takes time and effort. Due to its heterogeneous nature, the requirement for IoT devices is different. The NFV technology integrated with SDN can overcome this resource allocation problem and effectively handle network management.

6.3 Data Volume and Analysis

The use of IoT has risen due to the availability of high-speed internet and low-cost IoT devices. These IoT devices produce a large volume of heterogeneous data. Storing and analyzing this big data is a problem for industries nowadays. Cloudlet technologies such as cloud, fog, and edge computing techniques integrated with SDIoT can solve this big data issue. Cloud computing provides long-time data storage facilities and complex analysis techniques. The fog and edge computing provide temporary storage and immediate analysis results. It will make decision-making faster and reduce delays.

7. Conclusion

IoT is a promising solution for many automated applications. The integration with IoT and SDN provides flexibility and programmability to the network. In this chapter, we focused on SDN-based IoT applications and their architectures to offer QoS, cost-effective and reliable services to users. Moreover, different integrating technologies of SDIoT are discussed—blockchain and NFV. We identified some issues and challenges presented in the SDIoT network and highlighted different valuable approaches to address the challenges. To sum up, the integration of SDN and IoT is envisioned to help evolve scalable, energy-efficient, and cost-effective IoT architecture.

References

[1] Al-Fuqaha, A., Guizani, M., Mohammadi, M., Aledhari, M. and Ayyash, M. 2015. Internet of Things: A survey on enabling technologies, protocols, and applications. IEEE Communications Surveys Tutorials 17(4): 2347–2376.

[2] Chettri, L. and Bera, R. 2020. A comprehensive survey on Internet of Things (IoT) toward 5G wireless systems. IEEE Internet of Things Journal 7(1): 16–32.

[3] Swamy, S.N. and Kota, S.R. 2020. An empirical study on system level aspects of Internet of Things (IoT). IEEE Access 8: 188082–188134.

[4] Sadawi, A.A., Hassan, M.S. and Ndiaye, M. 2021. A survey on the integration of blockchain with IoT to enhance performance and eliminate challenges. IEEE Access 9: 54478–54497.

[5] Lohiya, R. and Thakkar, A. 2021. Application domains, evaluation data sets, and research challenges of IoT: A systematic review. IEEE Internet of Things Journal 8(11): 8774–8798. 10.1109/JIOT.2020.3048439.

[6] Saovapakhiran, B., Naruephiphat, W., Charnsripinyo, C., Baydere, S. and O˝ zdemir, S. 2021. QoE-Driven IoT Architecture: A comprehensive review on system and resource management. IEEE Access 10: 84579–84621.

[7] Lopes, F.A., Santos, M., Fidalgo, R. and Fernandes, S. 2016. A software engineering perspective on SDN programmability. IEEE Communications Surveys Tutorials 18(2): 1255–1272.

[8] Ali, J., Roh, B.H., Lee, B., Oh, J. and Adil, M. 2020. October. A machine learning framework for prevention of software-defined networking controller from DDoS attacks and dimensionality reduction of big data. pp. 515–519. In 2020 International Conference on Information and Communication Technology Convergence (ICTC). IEEE.

[9] Ali, J., Lee, B., Oh, J., Lee, J. and Roh, B.H. 2021. A novel features prioritization mechanism for controllers in software-defined networking. Comput. Mater. Contin. 69: 267–282.

[10] Sinche, S. et al. 2020. A survey of IoT management protocols and frameworks. IEEE Communications Surveys Tutorials 22(2): 1168–1190.

[11] Chen, S., Xu, H., Liu, D., Hu, B. and Wang, H. 2014. A vision of IoT: Applications, challenges, and opportunities with China perspective. IEEE Internet of Things Journal 1(4): 349–359.

[12] Siddiqui, S. et al. 2022. Toward software-defined networking-based IoT frameworks: A systematic literature review, taxonomy, open challenges and prospects. IEEE Access 10: 70850–70901.

[13] Theodorou, T. and Mamatas, L. 2017. CORAL-SDN: A software-defined networking solution for the Internet of Things. pp. 1–2. 2017 IEEE Conference on Network Function Virtualization and Software Defined Networks (NFV-SDN), Berlin, Germany.

[14] Jim´enez, M.B., Fern´andez, D., Rivadeneira, J.E., Bellido, L. and C´ardenas, A. 2021. A survey of the main security issues and solutions for the SDN architecture. IEEE Access 9: 122016–122038.

[15] Ali, J. and Roh, B.H. 2022. A novel scheme for controller selection in software-defined Internet-of-Things (SD-IoT). Sensors 22(9): 3591.

[16] Ali, J. and Roh, B.H. 2022. An effective approach for controller placement in software-defined Internet-of-Things (SD-IoT). Sensors. 2022 Apr 13; 22(8): 2992.

[17] Sahrish Khan Tayyaba, Munam Ali Shah, Omair Ahmad Khan and Abdul Wahab Ahmed. 2017. Software Defined Network (SDN) Based Internet of Things (IoT): A road ahead. pp. 1–8. In Proceedings of the International Conference on Future Networks and Distributed Systems (ICFNDS '17). Association for Computing Machinery, New York, NY, USA, Article 15.

[18] Rahouti, M., Xiong, K. and Xin, Y. 2021. Secure software-defined networking communication systems for smart cities: current status, challenges, and trends. *In*: IEEE Access 9: 12083–12113, doi: 10.1109/ACCESS.2020.3047996.

[19] Islam, M.J. et al. 2022. Blockchain-SDN-Based energy-aware and distributed secure architecture for IoT in smart cities. In IEEE Internet of Things Journal 9(5): 3850–3864. doi: 10.1109/JIOT.2021.3100797.

[20] Chakrabarty, S. and Engels, D.W. 2016. A secure IoT architecture for Smart Cities. pp. 812–813. 2016 13th IEEE Annual Consumer Communications Networking Conference (CCNC), Las Vegas, NV, USA. doi: 10.1109/CCNC.2016.7444889.

[21] Han, S.H. and Park, S.K. 2011. Performance analysis of wireless body area network in indoor off-body communication. In IEEE Transactions on Consumer Electronics 57(2): 335–338. doi: 10.1109/TCE.2011.5955164.

[22] Roy, C., Saha, R., Misra, S. and Niyato, D. 2022. Soft-health: software-defined fog architecture for IoT applications in healthcare. In IEEE Internet of Things Journal 9(3): 2455–2462. doi: 10.1109/JIOT.2021.3097554.

[23] Baktir, A.C., Tunca, C., Ozgovde, A., Salur, G. and Ersoy, C. 2018. SDN-based multi-tier computing and communication architecture for pervasive healthcare. In IEEE Access 6: 56765–56781. doi: 10.1109/ACCESS.2018.2873907.

[24] Sarkar, J.L. et al. 2022. I-Health: SDN-Based fog architecture for IIoT applications in healthcare. In IEEE/ACM Transactions on Computational Biology and Bioinformatics. doi: 10.1109/TCBB.2022.3193918.

[25] Barka, E., Dahmane, S., Kerrache, C.A., Khayat, M. and Sallabi, F. 2021. STHM: A secured and trusted healthcare monitoring architecture using SDN and blockchain. Electronics 10: 1787.

[26] Fang, X., Misra, S., Xue, G. and Yang, D. 2012. Smart grid—the new and improved power grid: A survey. IEEE Communications Surveys Tutorials 14(4): 944–980, Fourth Quarter. doi: 10.1109/SURV.2011.101911.00087.

[27] Chekired, D.A., Khoukhi, L. and Mouftah, H.T. 2018. Decentralized cloud-SDN architecture in smart grid: A dynamic pricing model. IEEE Transactions on Industrial Informatics 14(3): 1220–1231. doi: 10.1109/TII.2017.2742147.

[28] Grammatikis, P.R., Sarigiannidis, P., Dalamagkas, C., Spyridis, Y., Lagkas, T., Efs-tathopoulos, G., Sesis, A., Pavon, I.L., Burgos, R.T., Diaz, R., Sarigiannidis, A., Papamartzi-vanos, D., Menesidou, S.A., Ledakis, G., Pasias, A., Kotsiopoulos, T., Drosou, A., Mavropou-los, O., Subirachs, A.C., Sola, P.P., Dom´ınguez-Garc´ıa, J.L., Escalante, M., Alberto, M.M., Caracuel, B., Ramos, F., Gkioulos, V., Katsikas, S., Bolstad, H.C., Archer, D.E., Paunovic, N., Gallart, R., Rokkas, T. and Arce, A. 2021. SDN-based resilient Smart Grid: The SDN-microSENSE Architecture. Digital 1: 173–187. https://doi.org/10.3390/digital1040013.

[29] Adbeb, T., Di, W. and Ibrar, M. 2020. Software-Defined Networking (SDN) based VANET Architecture: Mitigation of traffic congestion. International Journal of Advanced Computer Science and Applications (IJACSA) 11(3). http://dx.doi.org/10.14569/IJACSA.2020.0110388.

[30] Islam, M.M., Khan, M.T.R., Saad, M.M. and Kim, D. 2021. Software-defined vehicular network (SDVN): A survey on architecture and routing, Journal of Systems Architecture, Volume 114.

[31] Helgason, K.S., Iversen, K. and Julca, A. 2021. Circular agriculture for sustainable rural development. https://www.un.org/development/desa/dpad/publication/un-desa-policy-brief-105-circular-agriculture-for-sustainable-rural-development/.

[32] Friha, O., Ferrag, M.A., Shu, L., Maglaras, L. and Wang, X. 2021. Internet of Things for the future of smart agriculture: a comprehensive survey of emerging technologies. In IEEE/CAA Journal of Automatica Sinica 8(4): 718–752. doi: 10.1109/JAS.2021.1003925.

[33] Hossain, M.S., Rahman, M.H., Rahman, M.S., Hosen, A.S.M.S., Seo, C. and Cho, G.H. 1987. Intellectual property theft protection in IoT based precision agriculture using SDN. Electronics 2021, 10: 1987.

[34] Sharma, P.K., Park, J.H., Jeong, Y.S. et al. 2019. SHSec: SDN based secure smart home network architecture for Internet of Things. Mobile Netw. Appl. 24: 913–924.

[35] Alonazi, W.A., HAMDI, H., Azim, N.A. and Abd El-Aziz, A.A. 2022. SDN architecture for smart homes security with machine learning and deep learning. International Journal of

Advanced Computer Science and Applications (IJACSA) 13(10). http://dx.doi.org/10.14569/IJACSA.2022.01310108.

[36] Ahvar, E., Ahvar, S., Raza, S.M., Vilchez, J.M.S. and Lee, G.M. 2021. Next Generation of SDN in Cloud-Fog for 5G and Beyond-Enabled Applications: Opportunities and Challenges. Network 1: 28–49. https://doi.org/10.3390/network1010004.

[37] Farris, I., Taleb, T., Khettab, Y., and Song, J. 2019. A Survey on Emerging SDN and NFV Security Mechanisms for IoT Systems. In IEEE Communications Surveys Tutorials 21(1): 812–837, Firstquarter. doi: 10.1109/COMST.2018.2862350.

[38] ABBASSI, Y. and Benlahmer, H. 2022. BCSDN-IoT: Towards an IoT security architecture based on SDN and blockchain. International Journal of Electrical and Computer Engineering Systems 2022-02-28. DOI: https://doi.org/10.32985/ijeces.13.2.8.

[39] Ninikrishna, T. et al. 2017. Software defined IoT: Issues and challenges. pp. 723–726. 2017 International Conference on Computing Methodologies and Communication (ICCMC), Erode, India, doi: 10.1109/ICCMC.2017.8282560.

[40] Manguri, K.H. and Omer, S.M. 2022. SDN for IoT Environment: A survey and research challenges. ITM Web Conf. 42: 01005. DOI: 10.1051/itmconf/20224201005.

[41] Jazaeri, S.S., Jabbehdari, S., Asghari, P. et al. 2021. Edge computing in SDN-IoT networks: a systematic review of issues, challenges and solutions. Cluster Comput 24: 3187–3228. https://doi.org/10.1007/s10586-021-03311-6.

CHAPTER 3

Revolutionising HR through the Deployment of Blockchain Technology
A Bibliometric Review

Rukma Ramachandran[1,]* and *Vimal Babu*[2]

1. Introduction

Blockchain Technology (BT) and Data Analytics (DA) have been rapidly adopted in various fields as they provide transparent and secured data visibility, in the field of Management. Several organisations have begun using the technology for supply chain management. Walmart, for example, has adopted a Blockchain-based system to track the supply chain of mangoes and pork in China, which has helped improve efficiency and transparency [1]. Furthermore, BT is also being employed in the sphere of finance to allow safe and transparent transactions. JP Morgan, for example, has deployed a BT for cross-border payments, reducing processing time from several days to a few hours [2], thus improving payment efficiency.

In the field of human resource management (HRM), BT transforms various HRM processes—it can securely store employee records, such as payroll information, contracts, and performance data. This can help increase the security and transparency of the HRM processes. Furthermore, BT can also be utilised to conduct background checks on candidates during the

---

[1] Research Scholar, SRM University AP, Amaravati.
[2] Associate Professor, SRM University AP, Amaravati.
* Corresponding author: rukuma_r@srmap.edu.in

recruitment process [3]. For example, a BT-based system can be used to keep a candidate's educational qualifications and job history, which authorised people can access.

The use of DA in HRM has become increasingly common as organisations seek to leverage data-driven insights for more effective decision-making. The Society for Human Resource Management (SHRM) has reported that 71% of HR professionals use data analytics in their decision-making process, with the most common areas of application being recruitment and talent management [4].

For example, global companies such as Google and Intel, are using DA to identify and assess potential candidates. Google has implemented an algorithm that predicts job performance based on various factors, including education, work experience, and the candidate's responses to a series of questions [5]. Similarly, Intel uses DA to identify patterns in employee behaviour and assess employee engagement, which helps in the development of employee retention strategies [6].

The scope for the use of BT in HRM is competitive and is a new area of exploration. BT helps in the verification of a job candidate's credentials. This streamlines the hiring process, by keeping a protected and absolute record of a candidate's education, work experience, and other credentials. The BT-based recruitment platform BlockRecruit, allows candidates to upload their verified credentials onto a blockchain, making it easier for recruiters to verify the information [7].

Thus, BT and DA collectively contribute to the development of blockchain-based HRM systems. These systems help store and manage employee data securely, while also leveraging data analytics to provide insights into employee behaviour and engagement. For example, the HRM platform Chronobank uses BT to store employee data securely and uses data analytics to provide insights into employee performance, which can help managers make better decisions about promotions, bonuses, and other incentives [8].

Therefore, this chapter understands the gap of the need to conduct Bibliometric analysis to understand the progress and trend in the rise of literature in BT and DA, particularly HR analytics. The purpose of the chapter is to:

1. Identify and analyse the rising pattern of the use of BT and DA in the HRM of organisations.
2. Determine authors and articles in the field of HRM of organisations with respect to BT and DA.

3. Explore the co-authorship patterns in the literature on BT and DA in the HRM of organisations and identify the most impactful collaborations.
4. Analyse the geographical distribution of research in HRM of organisations with respect to BT and DA and identify the leading countries in terms of research output.

2. Literature Survey

2.1 HRM in the Era of Emerging Technologies

Organisations are adopting new technologies to improve their HR functions, enhance efficiency, and streamline their processes. Recently, the rise of DA and BT has revolutionised HR management. DA has enabled HR departments to make data-driven decisions, while BT provides a safe and apparent platform for managing HR records. HR teams now gain insights into employee performance, engagement, and retention. As a result, they have been able to devise methods to boost employee performance and retention. According to [9], HR departments use DA to evaluate employee performance and make informed decisions about hiring, training, and development.

Blockchain technology is also being used to improve HR management. It provides a secure and transparent platform for managing HR records, such as employee contracts, performance evaluations, and payroll. A study by [10] found that BT is being used to enhance the security and efficiency of HR processes, such as background checks, CV verification, and onboarding.

In addition to DA and BT, other developing technologies, are also being used in HR management. For example, Artificial Intelligence (AI)-based recruitment tools are being used to screen job applicants and select the best prospects for the job. These technologies utilise machine learning algorithms to analyse resumes, cover letters, and other application materials, allowing HR teams to manage resources more efficiently and effectively during the recruitment process.

One example of an AI-based recruitment tool is Mya, an AI-powered chatbot that conducts initial candidate screening for companies such as Deloitte and L'Oréal. The chatbot asks relevant questions and assesses candidates' eligibility for the post, using natural language processing and machine learning. Another example is Pymetrics, a game-based assessment platform that uses neuroscience and AI to recruit the best candidates for a given job.

2.2 The Use of BT in HRM

Research studies have highlighted the possibility of BT in HRM. A study by Wang et al. [10] found that BT is being used to enhance the security and efficiency of HR processes, such as background checks, C.V. verification and onboarding. The authors found that BT can provide a secure and tamper-proof platform for storing employee records, making it easier for HR departments to verify employee credentials and ensure compliance.

Similarly, another study [11], showcased blockchain's potential in talent management. The authors proposed a BT-based computer system for talent management. The system uses BT to store candidate information, verify credentials, and track the hiring process, making it easier for HR departments to manage and track talent acquisition and work on smart contracts to automate the recruitment and hiring process. As BT's use grows in organisations, more HR departments are anticipated to utilise blockchain-based tools to optimise their HR operations.

2.3 Use of Data Analytics in Human Resource Management

Data analytics has revolutionised HRM by enabling HR departments to **make data-driven decisions.** Vast amounts of data on employee performance, productivity, and engagement can now be analysed leveraging DA, enabling the department to analyse patterns, trends, and insights that can be used to improve HRM practices.

HR departments, for example, can use predictive analytics to **discover the best candidates for job openings** by analysing resumes, career history, and competencies. Analytics can also be used to assess employee performance and find areas for improvement. According to Wang et al. [12], DA is being utilised in HRM to boost employee engagement and retention by analysing employee feedback, performance statistics, and social media data.

Data analytics can also be used to **track employee well-being and mental health**. DA can be used to detect early indicators of employee burnout and stress by analysing employee behaviour, communication patterns, and social media activity [13]. This can assist HR departments in developing ways to enhance employee well-being and lower the risk of burnout.

Furthermore, DA can be utilised **to assess the efficacy of HRM programmes and policies**. HR departments can track the effectiveness of training and development programmes, employee perks, and other HR activities. This can assist HR departments in making informed judgments about future HRM strategies and investments.

Therefore, with the use of DA in HRM it is becoming increasingly important for organisations to stay competitive and improve their HRM practices. By leveraging data analytics, HR departments can gain valuable insights into employee performance, engagement, and well-being, and use this information to make informed decisions about recruitment, retention, training, and development.

2.4 Collective Contribution of Blockchain Technology and Data Analytics in Human Resource Management

Blockchain technology and Data Analytics can collectively improve HRM practices in organisations. Blockchain provides a secure and transparent platform for managing HR records, while DA can be used to identify and analyse the hash function in the chain of blocks. This can enable HR departments to improve their decision-making processes, reduce costs, and increase efficiency.

For example, BT can be used to store employee records such as performance data, payroll information, and contracts. These records can be accessed by authorised personnel, and the records are tamper-proof, ensuring their integrity. DA can then be used to evaluate this data to identify trends, patterns, and insights that can be used to improve HRM practices.

A study by Salah et al. [16] found that the combination of BT and DA can be used to create a secure and decentralised HRM system. The study projected a BT-based HRM system that uses data analytics to analyse employee data and make informed decisions about HRM practices.

3. Methodology

BT has emerged recently due to its potential to revolutionise various industries. In the field of HRM, decision-making, and analytics, BT has been suggested as a solution to various problems such as data privacy, security, and transparency. However, a systematic literature review related to the use of BT in these areas is still lacking. Thus, the study aims to identify and screen articles related to the use of BT and DA in HRM, decision-making, and provide an overview of the research conducted in this area.

For the accomplishment of the purpose of this study, we used the **Preferred Reporting Items for Systematic Reviews and Meta-Analyses (PRISMA)** protocol. The PRISMA protocol is a widely recognised framework for conducting systematic reviews and meta-analyses. It consists of four stages: identification, screening, eligibility, and inclusion. Table 1 shows the PRISMA protocol as explained below:

3.1 Identification

We conducted a comprehensive search of an electronic database (Scopus) for articles related to the use of BT in HRM, decision-making, and analytics. The search was conducted on 10th April 2023, and we used the following keywords: "blockchain" AND ("human resource" OR "HRM" OR "analytics" OR "decision making"). The search strategy was designed to identify articles

Table 1. PRISMA flow diagram.

Stage	Total articles	Excluded articles	Included articles
Identification	2598	-	-
Screening	1108	530	578
Eligibility	1070	33	1037
Inclusion	503	534	-

that were relevant to our research question. The total number of articles retrieved in this stage was 2598.

3.2 Screening

After conducting the initial search, the following criteria were applied to narrow down the articles to be included in our study.

- the search was limited to articles published between 2017 and 2022.
- only open-access manuscripts from the subjects of Computer Science, Decision Science, Business Management, Accounting, Social Science and, Arts & Humanities were included.
- only articles, book chapters, and review documents were included.

After applying these criteria, 1108 articles remained.

3.3 Eligibility

Only quality journal articles were included in our study, and duplicate entries, incomplete records, and documents with less than 5 citations were excluded. We conducted this process manually by comparing the titles, authors, and publication information of the articles. After applying these criteria, 1070 articles remained.

3.4 Inclusion

Finally, we reviewed the abstracts and full texts of the remaining articles to determine whether they fell within the scope of our study. We included documents that were related to BT in HRM, decision-making, and analytics. The final number of articles that met the inclusion criteria was 503.

We used the PRISMA protocol to identify and screen articles related to BT in HRM, decision-making, and analytics. Our systematic review process allowed us to identify 2598 articles in the initial search, which was reduced to 503 after applying inclusion criteria. This report presents the methodology we used and provides statistics related to the identification, screening, eligibility, and inclusion of articles. By using a rigorous methodology, we were able to provide a comprehensive overview of the literature BT and DA in HRM.

4. Results and Findings

4.1 Current Trends

The statistics in Table 2 show an increase in the publication of articles on both BT and DA in recent years. In 2017, there were only two publications on the topic, but by 2022, there were 190 papers published, indicating the increasing interest in these emerging technologies.

The rise in publications on BT and DA suggests that both these technologies have significant potential in HR management. BT provides a secure and transparent platform for managing HR records, which can reduce fraudulent actions, errors, and data ethics. The use of BT in HR can improve the security and efficiency of HR processes, such as background checks, CV verification, and payroll.

Table 2. Trend of publications.

Year of publication	Number of papers
2017	2
2018	21
2019	47
2020	99
2021	145
2022	190

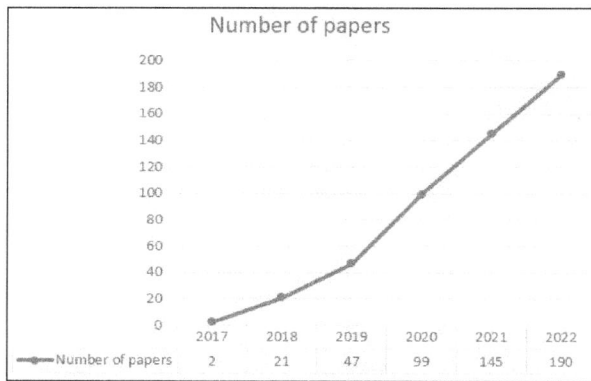

Figure 1. Trends showing the number of papers published.

DA, on the other hand, enables HR departments to leverage big data analytics to analyse the patterns, trends, and perceptions that can be used to improve HRM practices. Data analytics is being used in several HRM areas, including recruitment, employee retention, and training and development.

Overall, the increasing number of publications on Blockchain Technology and Data Analytics suggests that these technologies are becoming more important in HR management. The potential benefits of these technologies include improved efficiency, better decision-making, and enhanced security. As more organisations adopt these technologies, it is persistent to play an increasingly important role in the future of HR management.

4.2 Most Influential Authors and Publications—Citation Analysis

Citation analysis is an essential part of Bibliometric analysis and involves the identification of the number of occurrences of articles that have been cited by other works. This helps researchers to analyse the most influential publications and authors in a particular field of study, as well as the relationships between different research topics. This information can be used to understand the evolution of research trends over time, to identify research gaps, and to highlight areas where further research is needed.

VOS Viewer is a popular software tool used for citation analysis. The tool is designed to help researchers visualise Bibliometric networks, including co-authorship and citation networks. It provides an easy way to explore large bibliographic datasets and identify key publications, authors, and research topics. The current study utilises this software to serve the research objectives.

The aim of conducting this analysis in Bibliometric research is fulfilled with deeper insights into the research landscape in a particular field. By analysing citation patterns, researchers can identify the most influential works and authors, as well as the relationships between different research topics. Therefore, to uncover the facts, and to lay future research directions, the study aims to achieve its purpose. Overall, it is an important measurement for obtaining interpretation into the research avenue and can help researchers make more informed decisions about their research priorities and strategies.

The analysis in Table 3, uncovers the top authors in the area of HRM of organisations with respect to BT and DA and is dominated by authors who have researched and published works on different areas of BT [14].

Most of the articles in the list focus on the application of BT in industrial and managerial functions. The articles also explore the use of BT to address issues such as transparency, traceability, and sustainability in Industry and Management.

The authors on the list are from various parts of the world, with a significant number of authors from India and China. The list includes authors from both

Table 3. Most influential authors and articles.

References	Citations
[14]	704
(Dai et al. 2019) [15]	564
[16]	411
[17]	372
[18]	371
[19]	354
(Kouhizadeh et al. 2021) [20]	348
[18]	340
[21]	273
[22]	251
(Bai and Sarkis 2020) [23]	246
[24]	225
(Kang et al. 2020) [25]	223
(Sharma et al. 2020) [26]	205
(Xu et al. 2020) [27]	193
(Zheng et al. 2019) [28]	192
(Ahad et al. 2020) [29]	186
(Rahman et al. 2019) [30]	181
(Alladi et al. 2019) [31]	180
[32]	180

academia and industry, highlighting the importance of collaboration between academia and industry in researching and developing blockchain technology applications.

4.3 The Most Impactful Collaboration of Co-Authorship

Analysing co-authorship of documents based on organisations is an important aspect of Bibliometric analysis. This analysis provides insights into the collaborative research efforts of organisations and helps identify the most productive and influential organisations in a given field.

Table 4 shows the analysis of co-authorship of documents based on organisations which are listed along with number of citations and strengths. The organisations listed have published three to four documents and have received citations ranging from 57 to 370. The total link strength is a measure of the strength of the co-authorship links between the authors from the same organisation. It provides information on how closely the authors are connected in terms of their research collaborations. An organisation with a higher link

Table 4. Citations of co-authorship of documents based on organisation.

Organisation	Documents	Citations	Total Link Strength
Nanyang Technological University, Singapore, Singapore	3	370	1
California State University, Bakersfield, United States	3	314	1
University Of Texas at San Antonio, San Antonio, United States	3	251	1
Aston University, Birmingham, United Kingdom	4	240	0
International Institute of Information Technology, Hyderabad, India	3	88	0
Gumushane University, Gumushane, Turkey	3	57	1

strength indicates a stronger collaboration among its authors, which can lead to more impactful research outputs.

4.4 Leading Countries in Terms of Research Output

Bibliographic coupling is a Bibliometric analysis method, that compares the number of common references shared by two papers to determine how similar their research interests are. It is an essential research method for establishing links between research papers and authors.

In the context of BT and DA, Bibliographic coupling can help identify the leading countries in terms of research contribution towards this field. By analysing the count of citations, documents, and total link strength of different countries, we can understand the level of research activities and collaborations between countries in this field.

Table 5 gives an insight into the Bibliographic coupling of leading countries in terms of research contribution towards BT and DA. The Table shows the country (region), along with the total count of documents, citations, and strength.

The United States, China, and India have the highest number of documents and citations, indicating their leading contribution towards research on BT and DA. The same is shown in Fig. 2. Of importance is that, the total link strength of India is higher than that of China and the United States, indicating a higher degree of collaboration and interconnectivity between Indian research papers on this topic. Other countries, such as the United Kingdom, Australia, and South Korea, also have made significant contributions to research in this field. These insights can help researchers and policymakers identify the leading countries and potential collaborators in this field, which can lead to better cooperation and research outcomes.

Table 5. Bibliographic coupling of countries.

Region	Citations	Documents	Total Link Strength
United States	6123	112	259
China	5452	132	208
India	3745	99	209
United Kingdom	2687	62	171
France	1873	21	64
Australia	1841	47	94
Germany	1418	23	49
South Korea	1007	35	98
Russian Federation	1005	10	56
New Zealand	857	8	50
United Arab Emirates	818	17	49
Pakistan	768	17	72
Saudi Arabia	754	30	35
Taiwan	670	23	57
Canada	637	29	57
Italy	584	17	43
Turkey	500	25	67
Finland	446	11	38
Iran	381	14	45
Sweden	365	10	37

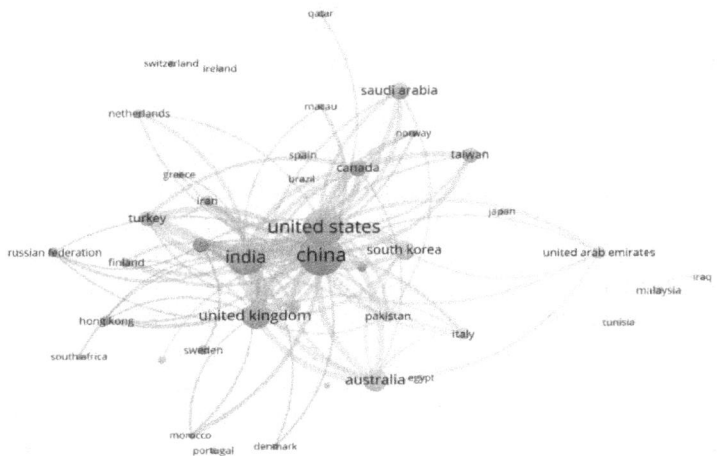

Figure 2. Bibliographic coupling of countries.

4.5 Co-citation Analysis

The frequency of co-citation of references in other publications is used to determine the strength of links between documents in Co-citation Analysis. It is a widely used approach for identifying influential works and writers, as well as for mapping the intellectual structure of a research topic. Co-citation analysis includes author co-citation, journal/source co-citation, and keyword co-citation. The examination of author co-citations requires determining the frequency with which writers are mentioned in other works jointly. The study of journal/source co-citations involves determining how frequently journals or sources are mentioned together in other publications. Keyword co-citation analysis is the technique of determining the frequency with which keywords are mentioned together in other publications.

The Process

There are two main methods of Co-citation Analysis:-**full counting and fractional counting.**

Full counting involves assigning each co-citation a value of 1, regardless of the number of co-citations involving that pair of documents. This method is simpler and easier to apply, but it may overestimate the strength of relationships between documents.

Fractional counting involves assigning a fraction of a co-citation to each document and documents cited in the co-citation. Although it is more complex, but is considered to be more accurate, as it accounts for the number of co-citations involving each document. In the study mentioned, co-citation analysis was performed on authorship co-citation and journal/source co-citation using fractional counting methods. Authorship co-citation analysis involved identifying the frequency with which authors were cited together in other publications related to the research field of interest. Journal/source co-citation analysis involved identifying the frequency with which journals or sources were cited together in other publications related to the research field of interest. The fractional counting method was used to account for the number of co-citations involving each document, which provided a more accurate representation of the strength of relationships between documents. Co-citation analysis is a useful technique for identifying important works and authors in a research field and mapping the intellectual structure of that field.

Table 6 represents the co-citation analysis of authorship co-citation. This analysis aims to identify the most cited authors within a particular research field. In this case, the focus is on BT and DA. The Table above provides information on the count of citations and strength for each author. Citations refer to the count of an author's work that is cited by different authors, while

Table 6. Co-citation of authorship.

Author	Citations	Total Link Strength
Zhang, Y.	328	272.08
Sarkis, J.	283	193.91
Wang, Y.	232	212.48
Wang, H.	222	198.55
Wang, J.	221	203.69
Wang, X.	190	174.99
Gunasekaran, A.	188	140.9
Kumar, N.	185	124.86
Chen, X.	183	169.59
Liu, Y.	174	159.64
Nakamoto, S.	160	133.72
Zhang, J.	159	150.05
Kouhizadeh, M.	153	137.28
Li, Z.	151	140.05
Li, J.	147	136.07
Li, Y.	146	132.93
Zhang, X.	146	135.5
Zheng, Z.	133	123.4
Chen, Y.	126	114.49
Li, X.	124	111.69

total link strength refers to the sum of all the links between the given author and other authors within the same co-citation network.

From Fig. 3 below, we can observe that Zhang, Y. has the highest number of citations and total link strength, indicating that his work is frequently cited by other authors in the field. Similarly, Sarkis, J. and Wang, Y. also have high citations and link strength, suggesting that their research is influential in the field. Other notable authors in Fig. 3 include Gunasekaran, A., Kumar, N., and Kouhizadeh, M., who have the highest count of citations and total link strength, indicating that their research is influential and widely cited by other authors in the field.

Overall, this Co-citation Analysis of authorship co-citation provides insights into the most influential authors in the field of BT and DA, which can analyse the rising research pattern and scope of future research.

Table 7 shows the co-citation analysis of sources, which refers to the analysis of the frequency of the co-citation of sources in the bibliography of

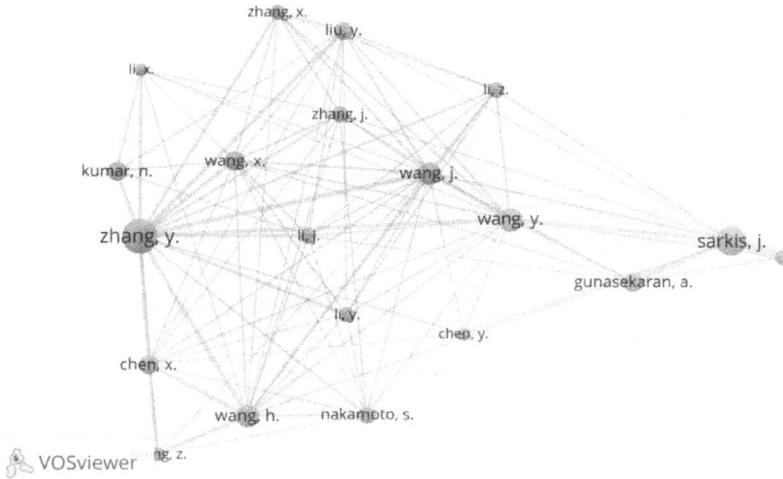

Figure 3. Co-citation analysis of cited authors.

Table 7. Co-citation of fractional cited sources.

Source	Citations	Total Link Strength
IEEE Access	881	471.49
International Journal of Production Research	345	178.14
Sustainability	318	222.13
Sensors	239	167.74
IEEE Internet Things Journal	218	127.74
Journal Of Cleaner Production	205	146.81
International Journal of Production Economics	192	143.8
International Journal of Information Management	190	128.35
International Journal Production Economics	175	113.35
Technological Forecasting and Social Change	169	73.79
International Journal Production Research	159	114.65
Journal Cleaner Production	149	105.72
IEEE Internet Things Journal	145	94.48
Technological Forecasting and Social Change	138	90.23
Future Generation Computer Systems	122	86.62
Journal Cleaner Production	113	50.54
Expert Systems with Applications	89	69.91
Computers & Industrial Engineering	84	62.36
IEEE Transaction on Industrial Informatics	84	61.52
Future Generation Computer Systems	80	66.38

articles. The sources listed in the Table have been frequently cited in articles related to BT and DA.

The source with the maximum citations and strength is IEEE Access, with 881 citations and a total link strength of 471.49. This indicates that IEEE Access is the most influential source in the field of blockchain technology and data analytics. The International Journal of Production Research comes in second with 345 citations and a total link strength of 178.14, followed by Sustainability with 318 citations and a strength of 222.13. It is also interesting to note that IEEE Internet Things Journal appears twice in the Table above, indicating its high relevance in the field. Journal of Cleaner Production and Technological Forecasting and Social Change also appear twice in the Table, indicating their continued importance in the field.

The Co-citation Analysis of sources shown in Fig. 4, provides insights into the most influential sources in a field, as well as the relationships between the sources. This information can be useful for researchers in identifying key works and developing research strategies. The fractional counting method used in this analysis provides a more accurate representation of the influence of sources by taking into account the count of sources cited and the count of sources that it co-cites with.

Figure 4. Co-citation of fractional cited sources.

5. Discussion

Bibliometric analysis is an important tool of research which is used to analyse the impact and influence of scholarly publications in a particular field of study. In recent times, the use of BT and DA in HRM have gained much attention from scholars and practitioners alike. As a result, it becomes critical

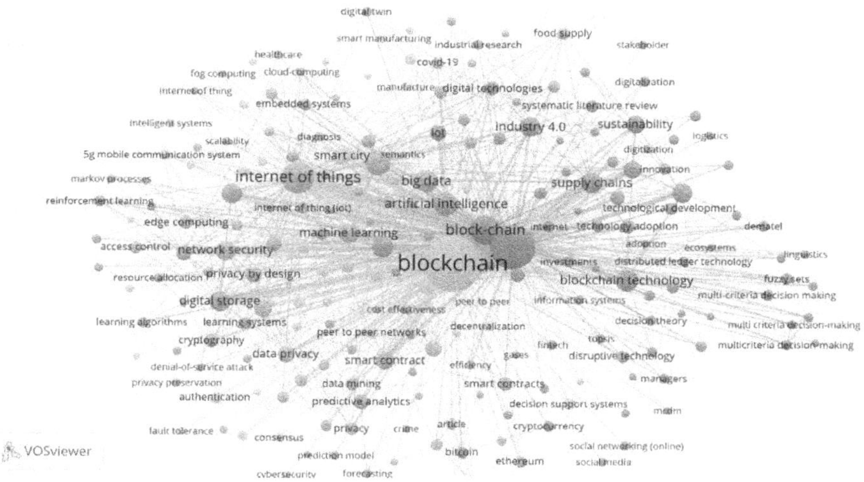

Figure 5. Co-occurrence of keywords.

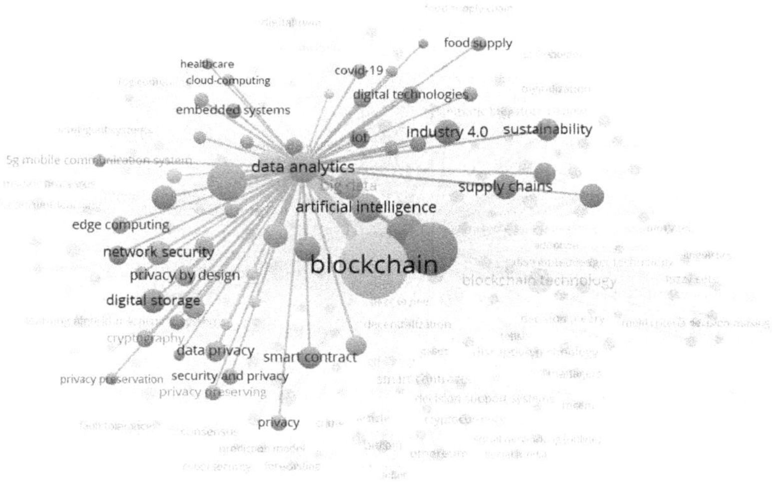

Figure 6. Co-occurrence of data analytics and blockchain.

to consider and analyse the upcoming patterns in this area of study, to provide insights for future research and practice.

As evident in Figs. 5 and 6, one can easily identify the need to include BT and DA in the HRM of an organisation.. As seen in the figures, a negligible number of sub-areas of HRM can be found. This attracts more publications in collaboration with the HRM of an organisation to work with the collaborative effort of BT and DA.

This study utilises Bibliometric Analysis to achieve the research objectives, which are to identify current trends, most influential authors and articles, co-authorship patterns, and leading countries in the use of BT and DA in HRM of organisations. The Bibliometric Analysis enables the identification and analysis of research patterns, collaborations, and contributions of scholars and institutions to a particular field of study.

This discussion will provide insights into the Bibliometric review from the study and their implications and directions for future practice. The use of Bibliometric analysis in this study provides a comprehensive understanding of the research landscape in the area of BT and DA in HRM of organisations. The findings of this study highlight the increasing trend in the use of BT and DA in the HRM of organisations.

The use of BT in HRM has gained significant attention due to its capability to enhance transparency, data privacy, and security in HRM processes. Similarly, the use of DA in HRM has been found to enhance decision-making, improve recruitment and selection processes, and increase employee engagement and retention.

The co-authorship analysis revealed that there is a significant level of collaboration in the literature on BT and DA in HRM, with Zhang, Y., Wang, Y., and Chen, X. being the most impactful collaborators. This implies that collaboration is essential for the advancement of knowledge in this area. Therefore, future research should encourage collaboration among authors, institutions, and countries to facilitate the development of knowledge in this field.

The geographical analysis revealed that the United States, China, and India are the leading countries in terms of research output in the area of HRM with respect to BT and DA. This implies that these countries have invested significantly in research in this field. However, there is a need for increased research from other countries to enhance the diversity of perspectives and approaches in this area of research.

Based on the findings of this study, it is clear that the use of BT and DA in organisational HRM is an area of research that requires additional investigation. Researchers, institutions, and governments should invest more in this subject to better understand how these technologies might be used to improve human resource management operations. Collaboration between authors, institutions, and countries should be encouraged in order to enhance knowledge in this field. Finally, future research should aim to widen the scope of viewpoints and approaches to this topic by including other countries in the geographical analysis.

6. Limitations and Future Scope of Research

- The study is based on a specific set of keywords and databases, which may not cover all relevant articles and authors in the field. It is possible that some relevant publications were excluded from the analysis due to the limitations of the search terms used.

- The study only considers articles written in English, which may have resulted in the exclusion of important research published in other languages.

- Bibliometric Analysis being a quantitative method relies on the availability and accuracy of citation data. However, citation data can be influenced by factors such as self-citation and citation biases, which may limit the accuracy of the analysis.

- Bibliometric Analysis cannot provide a complete picture of the research landscape and should be complemented with qualitative methods such as expert interviews and case studies to gain an inclusive understanding of the trends and practices in the field.

Overall, while the Bibliometric Analysis provides valuable insights into the present research on BT and DA in HRM of organisations, the findings should be interpreted with caution and considered in conjunction with other sources of information.

Based on the findings of this study, there is scope for future research that can be explored in the area of BT and DA in HRM. Firstly, while this study focused on the Bibliometric Analysis of publications in this field, future research can investigate the actual implementation and adoption of BT and DA in HRM practices. This can involve case studies and surveys of organisations to gain insights into the challenges, benefits, and outcomes of using these technologies. Secondly, the co-authorship analysis revealed the most impactful collaborations in this field. Future research can explore the nature and dynamics of these collaborations, and investigate how it leads to the development and knowledge in this area. Thirdly, the geographical distribution analysis identified the leading countries in terms of research output. Future research can investigate the factors that contribute to the research output of these countries, such as government policies, funding support, and academic culture. Fourthly, while this study focused on the use of BT and DA in HRM, future research can explore the potential of other emerging technologies, such as artificial intelligence, machine learning, and big data analytics, in enhancing human resource management practices.

7. Conclusion

This study conducted a Bibliometric Analysis to investigate and comprehend the use of BT and DA in organisational HRM. The study also aimed to identify the most influential authors and articles in this field, as well as to evaluate co-authorship patterns and the geographical distribution of research output. The Bibliometric Analysis revealed that the use of BT and DA in HRM is a rapidly growing research area. The most influential authors and articles were identified, and the co-authorship patterns showed that collaborations among researchers are essential for the advancement of this field. According to the geographical survey, the leading countries in terms of research production are the United States, China, and India. This study has some limitations, such as the exclusion of non-English language publications and the possibility of data errors in the databases. Nonetheless, this study contributes to the literature by offering insights into current trends as well as the most influential writers and papers in this field, and it can be utilised as a reference for future academics to further the research on BT and DA in organisational HRM.

References

[1] Kshetri, N. 2018. Blockchain's roles in meeting key supply chain management objectives. International Journal of Information Management 39: 80–89.

[2] Liao, Y. 2019. The development of sustainable supply chain management over time: A Bibliometric review. Journal of Cleaner Production 208: 1309–1318.

[3] Ratten, V. 2020. Coronavirus (Covid-19) and entrepreneurship: A review and research agenda. International Entrepreneurship and Management Journal 16(2): 293–323.

[4] SHRM. 2019. The business case for diversity, equity, and inclusion. Society for Human Resource Management. Retrieved from https://www.shrm.org/hr-today/trends-and-forecasting/research-and-surveys/pages/business-case-for-diversity-inclusion.aspx.

[5] Columbus, L. 2019. 82% of enterprises are prioritizing blockchain spending in 2019. Forbes. Retrieved from https://www.forbes.com/sites/louiscolumbus/2019/04/07/82-of-enterprises-are-prioritizing-blockchain-spending-in-2019/?sh=5d930eea4b4f.

[6] Nguyen, N.T. 2019. Effects of product diversification and coordination on supply chain performance in the presence of free-riding behaviour. International Journal of Production Economics 212: 98–114.

[7] Gorecki, P. 2018. Blockchain technology adoption: A diffusion of innovation theory perspective. Journal of Innovation Management 6(4): 28–48. doi: 10.24840/2183–0606_006.004_0002.

[8] Chronobank. (n.d.). Chronobank White Paper. Retrieved from https://chronobank.io/downloads/whitepaper.

[9] Olszewska, J.A. and Dyduch, J. 2019. Determinants of sustainable procurement in supply chains: Evidence from the literature. Sustainability 11(5): 1468.

[10] Wang, X., Zhang, Y. and Hou, L. 2020. Secure multi-user data sharing for edge computing in 5G and beyond networks. Journal of Network and Computer Applications 168: 102690.

[11] Chowdhury, A.R. and Alzoubi, K. 2020. Blockchain-based traceability in supply chain management: A comprehensive overview. Journal of Supply Chain Management, Logistics and Procurement 3(2): 238–262. doi: 10.1108/JOSCM–01–2020–0013.

[12] Wang, H., Liu, Y., Zhu, Y. and Zhang, D. 2018. Blockchain-based data integrity service framework for IoT data. Future Generation Computer Systems 87: 213–221.

[13] Albdour, A.A. and Altarawneh, I.I. 2021. The role of blockchain in supply chain management: A systematic review. Journal of Industrial Engineering and Management Science 1(1): 29–41.

[14] Ivanov, D., Dolgui, A and Sokolov, B. 2019. The impact of digital technology and Industry 4.0 on the ripple effect and supply chain risk analytics. International Journal of Production Research 57(3): 829–846. doi: 10.1080/00207543.2018.1463979.

[15] Dai, H.-N., Zheng, Z. and Zhang, Y. 2019. Blockchain-based internet of things: A systematic survey. Journal of Industrial Information Integration 17: 100120. doi: 10.1016/j.jii.2019.100120.

[16] Salah, K., Rehman, M.H.U., Nizamuddin, N. and Al-Fuqaha, A. 2019. Blockchain for AI: Review and open research challenges. IEEE Access 7: 10127–10149. doi:10.1109/ACCESS.2018.2890507.

[17] Zhang, P., White, J., Schmidt, D.C., Lenz, G. and Rosenbloom, S.T. 2018. FHIR Chain: Applying blockchain to securely and scalably share clinical data. Computational and Structural Biotechnology Journal 17: 1–7.

[18] Kamble, S.S., Gunasekaran, A. and Gawankar, S.A. 2020. An integrated framework for evaluating sustainable supply chain performance: Application of grey theory and decision-making trial and evaluation laboratory technique. Journal of Cleaner Production 270: 122259.

[19] Kang, J., Zehui, X., Dusit, N., Shengli, X. and Junshan, Z. 2019. Incentive mechanism for reliable federated learning: a joint optimization approach to combining reputation and contract theory. IEEE Internet of Things Journal 6(6): 10700–714. https://doi.org/10.1109/JIOT.2019.2940820.

[20] Kouhizadeh, M., Saberi, S. and Sarkis, J. 2021. Blockchain technology and the sustainable supply chain: Theoretically exploring adoption barriers. International Journal of Production Econonmics 231: 107831, Jan. 2021, doi: 10.1016/j.ijpe.2020.107831.

[21] Puthal, D., Malik, N., Mohanty, S.P., Kougianos, E. and , C. 2018. The blockchain as a decentralized security framework. IEEE Consumer Electronics Magazine 7(2): 18–21.

[22] Lezoche, M., Hernandez, J.E., Alemany Díaz, M. del M.E., Panetto, H. and Kacprzyk, J. 2020. Agri-food 4.0: A survey of the supply chains and technologies for the future agriculture. Comput. Ind. 117: 103187, May 2020, doi: 10.1016/j.compind.2020.103187.

[23] Bai, C. and Sarkis, J. 2020. A supply chain transparency and sustainability technology appraisal model for blockchain technology. International Journal of Production Research 58(7): 2142–62. https://doi.org/10.1080/00207543.2019.1708989.

[24] Fisch, C. (n.d.). 2019. Blockchain in healthcare: 14 benefits and challenges. Health IT Analytics. Retrieved from https://healthitanalytics.com/features/blockchain-in-healthcare-14-benefits-and-challenges.

[25] Kang, J., Xiong, Z., Niyato, D., Zou, Y., Zhang, Y. and Guizani, M. 2020. Reliable federated learning for mobile networks. IEEE Wirel. Commun. 27(2): 72–80, Apr. 2020, doi: 10.1109/MWC.001.1900119.

[26] Sharma, R., Kamble, S.S., Gunasekaran, A., Kumar, V. and Kumar, A. 2020. A systematic literature review on machine learning applications for sustainable agriculture supply chain performance. Computers & Operations Research 119: 104926, Jul. 2020, doi: 10.1016/j.cor.2020.104926.

[27] Xu, X., Zhang, X., Gao, H., Xue, Y., Qi, L. and Dou, W. 2020. BeCome: Blockchain-enabled computation offloading for IoT in mobile edge computing. IEEE Transactions on Industrial Informatics 16(6): 4187–4195, Jun. 2020, doi: 10.1109/TII.2019.2936869.

[28] Zheng, T., Ardolino, M., Bacchetti, A. and Perona, M. 2021. The applications of Industry 4.0 technologies in manufacturing context: a systematic literature review. International Journal of Production Research 59(6): 1922–1954, Mar. 2021, doi: 10.1080/00207543.2020.1824085.

[29] Ahad, M.A., Paiva, S., Tripathi, G. and Feroz, N. 2020. Enabling technologies and sustainable smart cities. Sustainable Cities Society 61: 102301, Oct. 2020, doi: 10.1016/j.scs.2020.102301.

[30] Rahman, M.A., Rashid, M.M., Shamim Hossain, M., Hassanain, E., Alhamid, M.F. and Guizani, M. 2019. IoT-based smart healthcare system: Architecture, implementation, and future research directions. Journal of Medical Systems 43(7): 180.

[31] Alladi, T., Chamola, V., Parizi, R.M. and Choo, K.-K.R. 2019. Blockchain applications for Industry 4.0 and industrial IoT: A review. IEEE Access 7: 176935–176951, doi: 10.1109/ACCESS.2019.2956748.

[32] Gao, F., Zhu, L., Shen, M., Sharif, Wan, Z. and Ren, K. 2018. A blockchain-based privacy-preserving payment mechanism for vehicle-to-grid networks. IEEE Network 32(6): 184–192, Nov. 2018, doi: 10.1109/MNET.2018.1700269.

CHAPTER 4

A Study on Health Monitoring of the Structures through Digital Twin

Galiveeti Poornima,[1,*] *Jagdesh H. Godihal*[2] and
Vinay Janardhanchari[3]

1. Introduction

According to Xu [107], the construction industry is commonly acknowledged as a sector that heavily relies on information, which requires precise, all-encompassing, and prompt communication in a clear format that the intended receiver can understand. During the complete duration of a construction project, a significant volume of information is generated, spanning from the preliminary ideation phase to the ultimate dismantling phase. Effective management of information flow, processing of vast amounts of data, and extraction of actionable insights are essential for achieving success in construction projects [20]. Scholars have observed that effective information management is critical at every construction project's lifecycle stage. This includes the creation and dissemination of information and its comprehension and application in the construction, upkeep, repurposing, and eventual disposal of the project. The academic discourse has predominantly centered on data management in the design and construction stages, despite acknowledging the importance of information throughout the project's lifespan. Despite the fact that these stages are crucial, they only contribute

[1] Asst. Prof. School of CSE.
[2] Professor, Civil Engineering Department, Presidency University.
[3] Cloud Operations, USA.
* Corresponding author: galiveetipoornima@presidencyuniversity.in

between 30 and 40% of the project's overall costs. According to Nical [75], the operating and usage phase comprises 60 to 80% of the entire project lifecycle.

The conservative stance of the construction industry towards technological advancements and their potential applications has been a subject of criticism. However, in recent decades, the industry has made significant strides in improving information management by implementing BIM [73]. 3D object information systems have replaced the conventional approach of the construction industry of utilizing 2D drawing information systems due to the advent of BIM [13]. BIM has been a pivotal instrument of innovation within the construction sector for over a decade. The attainment of objectives in building design, is contingent upon the adoption of a holistic approach, the promotion of collaboration and communication among key stakeholders, the enhancement of productivity, the improvement of the final product's overall quality, and the minimization of fragmentation. These factors have been identified as critical [90, 97]. As highlighted in reference [28], BIM offers a notable advantage in cohesively presenting project information throughout its entire life cycle, unlike a fragmented approach.

Industry 4.0, the fourth industrial revolution, employs the most recent advancements in Internet and Communication Technologies (ICT) to enable notable progress in diverse fields [63]. The principal aim of this initiative is to augment efficiency and efficacy by facilitating intercommunication between machines [106]. The industrial production technological advances of Industry 4.0, commonly known as "Digital Twins" (DT), are frequently characterized as a virtual copy of a physical asset with high precision and real-time bidirectional interaction capabilities for simulation objectives. Furthermore, it provides decision-supporting functionalities to improve product services. This definition is supported by reference [115]. DT provides a practical method for tracking resources, simulating scenarios, and coming up with solutions, and is frequently considered a flexible and scalable solution [100, 116]. There has been a surge of interest among academic and business communities in utilizing DT systems within the construction sector.

Although a significant study has been conducted on the utilization and advantages of BIM in the construction process, it has been noted that BIM must also capture the data produced during the operational and usage phases, as evidenced by previous studies [57]. Integrating innovative technologies is of utmost importance for the construction industry, especially in the context of the advent of Industry 4.0. The DT, a digitization technology, specifically developed to monitor a physical asset and enhance operational efficiency by acquiring real-time data that enables predictive maintenance and informs decision-making, constitutes a pivotal component of the Industry 4.0 strategy [57].

1.1 The Principle of Intelligent Building Construction

The notion of Integrated Construction (IC) is based on the integration of advanced construction technology [56, 72, 112, 114] and Information Technology (IT) [45, 47, 53, 60, 62], with a particular emphasis on Cyber-Physical Systems (CPS) [11, 30, 55], as reported by the Global Engineering Frontier 2018 [27]. Traditionally, the construction life-cycle, which includes project approval, design, manufacturing, transportation, assembly, operation and maintenance, and service, is when the activities related to making physical systems, like building parts, techniques, and components, are examined. The twin model of cyberspace [110, 113] is used to aid in the comprehension and utilization of data, the creation of models, the evaluation and forecasting of states, the application of intelligent optimization techniques, and the making of important decisions. This is achieved by applying knowledge concerning creating objects, processes, equipment, and procedures, encouraging innovation, and incorporating intelligent-sensing technology [49, 101, 111].

The construction industry has undergone significant technological advancements, prompting the introduction of Industry 4.0, which proposes a novel construction technology aimed at achieving high adaptability, rapid design changes through IT, and more flexible technical workforce training. CPS [11, 30, 55], BIM [36, 85], IoT [31, 91], big data [7, 9], and cloud computing [46, 108] are just some of the cutting-edge technologies currently used in construction. The notion of Industrial Ecology (IE) has gained heightened prominence in the era of Industry 4.0, owing to the growing urgency of sustainability as a critical issue. The concept of Intelligent Construction (IC) involves the integration of various technological advancements such as perception technology, data mining, design optimization, and management decision-making to ensure a smooth and efficient construction process. This approach aims to effectively coordinate all involved parties and achieve energy-efficient completion of construction tasks. These developments have been documented in various academic sources [33, 65, 78]. The phenomenon of diversification has been noted in the field of Integrated Circuits (IC), as there has been a proliferation of IC systems that are tailored to perform specific functions and are being utilised in real-world construction endeavors. As a result, there has been a significant increase in cognitive capacity in construction, as evidenced by studies [15, 83].

1.2 DT's Basic Idea

Greives [41] defines DT as "a model for creating an informational mirror image of a physical object" (animate and inanimate alike). Integrating physical and virtual realms facilitates uninterrupted data transmission and allows for the cohabitation of virtual and physical entities [34]. The definition of DT comprises two fundamental constituents. The initial emphasis is on elucidating the correlation between the tangible prototype and its digital simulation. To generate data in real-time, sensors [48, 74] establish a connection. As previously stated, DT refers to a digital representation of a physical entity that is updated in real time. It accurately maps physical objects, and depending on particular models, it can describe and improve these objects.

2. Review of Literature

This section analyzes the present status of Digital Twin implementation in the construction sector, focusing on its key features, specifically quick updates, and two-way coordination. Although smart buildings with automated systems are increasingly common, the construction sector has adopted DT less frequently than other industries. Digital Technology utilisation within the construction industry, possesses a vast scope that surpasses the mere incorporation of automated systems into structures. It also includes the synchronization of these systems, which makes it easier for the cyber and physical domains to work closely together. These systems are often called Cyber-Physical systems. Integrating these systems, in building design and construction presents opportunities for developing sophisticated and intelligent structures [10, 70]. These structures integrate automated sensors, actuators, and other technologies that act as a means of monitoring physical assets. This establishes a basis for the implementation of DT.

The functionality of DT is reliant on quick updates and bidirectional communication [80]. According to research, incorporating DT into the technological development process, provides a more adaptable and refined model, mainly when implemented in the early stages of product creation [43]. DT's advancement is contingent upon acquiring sensory data derived from the tangible prototype. The virtual model can efficiently manage and supervise the product by acquiring requisite data from tangible resources. According to reference [80], project managers and analysts can access the virtual model produced via their mobile devices. Thus, incorporating a DT during the initial stages of a project would guarantee precise synchronisation and efficient operation of the system.

The utilisation of DT technology enables the collection of data, which can facilitate the efficiency of construction activities and management, throughout the life cycle over an extended period. However, the extent of optimisation is dependent on the quantity of information gathered. The outcomes of DT are ascertained through the amalgamation of information and empirical evidence generated by the sensors integrated into the tangible systems. In terms of architecture and built environment,[1] DT has been determined to possess various attributes, including BIM and the ability to generate 3D and 2D representations, schedules, contracts, and construction documents such as submittals, change orders, and RFIs. Additionally, operational data obtained from embedded sensors and data derived from Artificial Intelligence (AI) and machine learning technology are also incorporated into DT. Integrating these characteristics enables intelligent building administrators to attain access, regulate, identify malfunctions, and make informed judgments concerning a property's diverse systems, workspaces, and conglomerates.

One example of implementing DT in the built environment is the Frasers Tower in Singapore. This establishment provides an interconnected collaborative workspace that caters to the needs of individuals and businesses, seeking to leverage DT for their operations.[2] Bentley Systems and Schneider Electric worked together on the project at hand. The project involved acquiring data through the utilization of 179 Bluetooth beacons in conference rooms and 900 sensors specifically intended for assessing lighting, temperature, and air quality. The platform above utilises embedded sensors and telemetry to generate 2,100 data points, subsequently linked to the cloud through Microsoft Azure. This integration facilitates the implementation of a comprehensive environmental management system. The implementation of digital technology solutions within the construction industry in Singapore needs to be improved in contrast to other sectors.

As per the findings of Bughin et al.'s research [26], the construction sector has a relatively low adoption rate of digitization, estimated at 1.4%. On the contrary, the ICT sector exhibits the most significant rate of digitization adoption, approximated at 4.6%. The investigation was carried out within the timeframe 2005 to 2014. Thus, the implementation of digitisation within the construction sector may have escalated after this period. Brilakis et al. [24] assert that, despite the potential benefits that DT technology can offer stakeholders in the built environment, its full potential has not yet been realized.

When developing a Decision Tree, it is imperative to define the objective for which it is being constructed clearly. The selection of the physical asset that is to undergo digitisation and the corresponding Level of Detail (LoD) necessary for the creation of the DT is dependent on the intended objective of the DT [24]. This is because a DT entails advanced analytical techniques and is a broad notion. DT can be defined as the application of digital technologies

---

[1] https://constructionblog.autodesk.com/digital-twin/

[2] https://www.pbctoday.co.uk/news/bim-news/digital-twin-smart-building/73253/

to achieve a project or industry objective within the construction domain. As per the findings of Braun et al. [23], the primary role of a DT is to monitor and oversee the advancement of construction projects. The DT objective to assess the condition of a bridge was defined by another application in the construction sector [88]. According to scholarly literature, a geometric DT is a computer-generated model augmented with semantic information, serving as the basis for constructing a comprehensive DT [24]. The first stage in creating a DT entails the production of a digital geometric representation that will be linked to the corresponding physical component serving as the foundation for the DT application.

BIM is a crucial instrument in a structure's planning, building, and maintenance. It can produce the necessary three-dimensional models that are necessary for the creation of a digital prototype [24]. BIM offers DT an essential input (including geometrical inputs), especially for new projects. It is noted that even though these processes are automated, they still take more time than using digital models created with BIM. Agapaki et al. [2] made this determination, identifying that the labor hours used to model 53,834 pipes were approximately 5,200, if the digital representations of the facilities were accessible, it is possible that the time spent could have been allocated toward other constructive endeavors. Nevertheless, endeavors are underway to employ Artificial Intelligence (AI) to automate the modeling of extant structures. The office space modeling was automated by Qi et al. [32] using semantic segmentation of points cloud deep learning. Apart from the ultimate chromatic manifestation, the output model maintains congruity with the input model about its geometry.

Brilakis et al. [24] have reported that geometric models are generated through Bottom-up and Top-down methodologies. The Bottom-up methodology involves the identification of geometric primitives, such as lines, planes, and cuboids, within a given point cloud. These primitives are subsequently organised and classified into more complex geometrical structures, and their spatial and functional interrelationships are established. Ultimately, a geometric model is generated based on the steps above. The Top-Down methodology produces geometric models by recognizing that items within the representations can be differentiated, based on their orientation and interdependence with other entities in the constructed setting, such as the interrelationship between towers, levels, and chambers. This approach is dependent on the asset's context, and it works well for buildings and overpasses that adhere to standardized contextual guidelines. The present discourse expounds on the techniques employed in generating geometric models from pre-existing structures, obviating the requirement of digital models to construct the requisite DT platform.

Following the creation of the digital representation of the DT, the subsequent step involves integrating it with the physical system. The bridge is equipped with sensors that gather pertinent information, which is subsequently modeled in the Internet of Things (IoT) cloud infrastructure to provide the expected instantaneous notifications. Utilisation of Artificial Intelligence (AI) and Machine Learning (ML) are employed to replicate the information derived from the tangible assets within the IoT cloud. DT must conduct sophisticated data analytics on the IoT cloud, to achieve contemporary updates and bidirectional communication. The functional specifications of the proposed DT platform serve as the basis for data analytics, as they delineate the platform's purpose. The physical entity undergoes uninterrupted data gathering through sensors accountable for transferring the data to the IoT platform, via the communication network. Following this, a data simulation is performed on the forum. The sensors are utilised to incorporate the feedback into the physical asset. The virtual model is subject to continuous updates contingent upon the physical asset's current condition.

3. Building Information Modeling

BIM has emerged as a crucial methodology for the digitization of the supply chain for the built environment. It is a knowledge-sharing methodology, that employs digital representations of a building's structural and functional elements [17]. Over recent years, the employment of incentives and BIM has experienced a notable increase. It has garnered significant attention due to its capacity to enhance quality and minimize costs and time [14]. Although BIM has been implemented in various significant design and construction undertakings, it has yet to become ubiquitous in the industry. Modern construction and design endeavors have exhibited enhanced efficacy in cost reduction, quality improvement, schedule adherence, and facilitation of superior communication among project stakeholders. Implementing BIM as a methodology has resulted in several benefits for project owners. These benefits include the potential to mitigate claims, simplify calculations and visualizations for marketing purposes, and promote interdisciplinary collaboration.

BIM originated during the 1950s and 1960s through Computer Aided Design (CAD). The development of CAD software can be traced back to 1963 when Ivan Sutherland created Sketchpad, a graphical user interface.

Subsequently, in the 1970s, the French Aerospace Company converted the software from two dimensions (2D) to three dimensions (3D). After this, Autodesk attained eminence in the realm of Information Technology throughout the 1980s and 1990s, owing to the triumph of its AutoCAD offering. The 4D model was introduced to aid stakeholders, particularly those in the Architecture, Engineering, Construction (AEC) sector, in effectively managing time-sensitive schedules and resource allocation. This model superseded the previous 3D model. Subsequently, the development of 5D was initiated in correlation with the estimation of project expenses. Quantity surveyors or cost estimators can utilize the 5D model to authenticate project costs during the estimation phase. The sixth dimension (6D) emphasizes sustainable practices more, whereas the seventh dimension (7D) is primarily concerned with managing facilities. Nevertheless, the functions above are imperative for the expansion of n-dimensional space. As per Beveridge's [18] findings, the 8D designation was intended for integrated project delivery and maintainability. Accordingly, the 9D assignment was allocated for acoustics, the 10D title was assigned for security, and the 11D appointment was reserved for heat. This illustrates that the development of BIM originated in the 1950s and has since undergone an ongoing evolution.

BIM is a swiftly growing and pioneering technology, widely employed in the global construction industry. This tool is imperative in modern engineering, construction, and architectural procedures. The technology above facilitates the generation of one or multiple precise digital representations of edifices, exhibiting commendable interoperability.

The ISO19650[3] standard endorses a novel technology that has been sanctioned by the government and is applicable to projects of varying levels and dimensions. By 2022, BIM softwares such as Revit, Navisworks, Tekla, BIM Collab, Plannerly, and Autodesk BIM 360 will enable interoperability. BIM encompasses various levels and dimensions. It has been utilized in various endeavors.

John Sisk & Son successfully integrated BIM into the Quintain Wembley project[4] in London by utilizing their Digital Project Delivery (DPD) methodology, as documented in a case study by BIM Plus. Furthermore, Revit was utilized to simulate the NHS Nightingale Hospital in East London.[5]

---

[3] https://www.bsigroup.com/en-GB/iso-19650-BIM/?
[4] https://www.bimplus.co.uk/case-study-sisk-scores-digitally-wembley/
[5] https://www.bimplus.co.uk/mace-turned-ai-tech-mayfair-project-meet-covid-19-/

3.1 Applications

The following are some of the ways in which different applications of a BIM can be put to use:

- Rendering in three dimensions can be quickly and easily produced in-house with very little additional labor required.
- Generating shop drawings for diverse building systems is a straightforward task. Upon completion of the model, the metal sheet ductwork shop drawing can be expeditiously generated. This serves as an illustration of the ease with which it can be accomplished.
- These models are available for use by fire departments along with other officials in the course of their evaluation of building projects.
- Modifying a Building Information Model to visually depict potential issues, such as leaks and evacuation plans, is straightforward.
- Facilities management departments can utilize BIM software to facilitate repairs, layout modifications, and maintenance operations.
- The BIM software comes equipped with a built-in cost feature estimation. Any changes made to the model cause automatic extraction and modification of the quantity of material.
- Material ordering, building component fabrication, and delivery can all be effectively scheduled with the help of a BIM.
- The three-dimensional scaling of BIM models, allows for comprehensive visual inspection of practical systems, thereby identifying and eliminating potential interferences. Using this procedure, it is possible to ensure that the pipes do not come into contact with any steel beams, ducts, or walls.

3.2 Benefits

As per the findings of CRC Construction Innovation [54], the primary benefit of BIM is its ability to offer a precise geometric depiction of a building's constituents in a unified data ecosystem. Additional advantages encompass the following aspects:

- It is much simpler to share information and is also possible to reuse it and add value to it.
- Analyzing building proposals, rapid simulation execution, and performance bench-marking are all viable methods for developing enhanced and innovative solutions.
- The cost of longevity is better understood, and there is greater predictability in environmental performance.

- The output of documentation is adaptable and makes use of automation.
- The electronic representation of product data can be utilized in subsequent stages of production, including the fabrication and integration of structural systems. In addition, this data can also be exploited for financial gains.
- When presented in a clear and concise manner, proposals are more easily understood.
- Data about prerequisites, blueprinting, erection, and functioning can all be employed in facilities administration.

4. DT-Key Concepts

4.1 History

The initial mention of a "twin" in the domain of DT can be traced to the aerospace industry and has its origins in NASA's Apollo project during the 1960s, as documented in NASA Technology Roadmaps [22, 73, 92]. The NASA project involved the construction of two spacecrafts, one of which would orbit the planet as 'the twin' and carry out the mission [22, 117]. Rosen et al. [86] employed the term "twin" as a representation that effectively mirrored the operational behaviour of the missions in real-time. According to Zhaung et al. [117], the "twin" was a tangible entity during that period. 'Twin' did not at this time include the digital component. The phrase "Digital Twin" was first introduced by Michael Grieves in 2003 within the context of his instruction on Product LManagement (PLM) [42]. PLM is a comprehensive approach that encompasses a range of business activities associated with data creation, modification, and utilization, spanning the entire product lifecycle from design and production to maintenance, recycling, and disposal [89]. According to Kritzinger et al. [61], DT is a digital representation of a physical system created as a distinct entity but continues to be linked to the biological system. This definition was derived from an industry presentation cited by the authors. Following definitions and implementations of DT have adopted this mode of reasoning, wherein intelligent devices are utilised to establish a connection between a digital representation and a tangible object, and the digital model is continuously updated in real time through a functional communication infrastructure. As per the findings of Grieves and Vickers [43], the DT should comprise comprehensive and relevant data about the physical system asset, which is acquired through rigorous examination of the real-world scenario. This is necessary to ensure that the DT accurately and effectively reflects the biological system. Figure 1 presents a concise overview of significant advancements in the evolution of the DT concept.

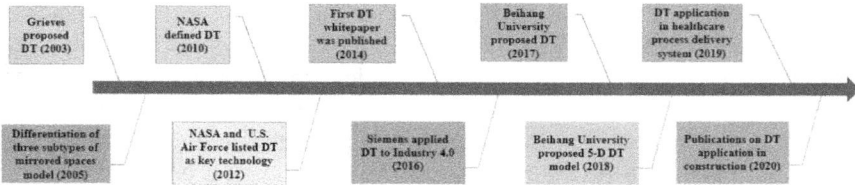

Figure 1. The DT milestone was developed as per the work of Qi et al. [60].

4.2 Definition of DT

The initial stage in comprehending the DT concept involves providing a lucidly articulated definition. Various scholars have proposed distinct interpretations to articulate the nature of DT technologies. Michael Grieves initially defined DT as an information mirroring model [44] during his Product Management course. Nonetheless, this definition is needed to comprehend the concept of DT comprehensively. Several authors have formulated the following definitions of DT. Rosen et al. [86] characterized DT as a fusion of tangible and intangible domains that are analogous to one another, to evaluate the life cycle functions of the fundamental component. This definition offers a concise elucidation of the role and components of the DT. Boschert and Rosen [22] subsequently provided a more concise definition of DT as encompassing all pertinent physical and functional information about a given system or product. The definition in question accentuated the significance of data interchange and the algorithms that regulate the behavior of both tangible and intangible models. Despite its brevity, this definition disregarded the constituent elements and objectives of DT data, instead opting to concentrate exclusively on the data itself. Grieves and Vickers [43] have expounded on the concept of DT in PLM, characterizing it as a collection of digital information constructs that precisely depict a physically manufactured product, encompassing its atomic and macro-geometric aspects. It is a digital depiction of the output. To gain a more comprehensive understanding of the digital design of the final product and to address any potential discrepancies in execution, it was proposed to conduct a comparison between a DT and its corresponding engineering design, which represents the physical manifestation of the object [43]. The current definition provides a more comprehensive account of DT compared to Grieve's report of 2005 [44], with a heightened focus on managing the life cycle.

According to Lui et al. [66], the concept of Digital Twin can be defined as a flexible depiction of a concrete resource or structure that adjusts promptly to operational changes, by gathering and evaluating real-time data and can anticipate future conditions of the related physical entity. Compared to previous definitions of DT, this specific definition demonstrates a higher level of comprehensiveness. A DT is a technological system that leverages various tools, technologies, and communication infrastructure (i.e., the internet) to collect data from tangible objects and replicate their digital representations. According to Madni et al. [68], the term DT refers to a digital portrayal of a system's maintenance, performance, and health status information continuously updated in real-time throughout the system's operational cycle. The definitions put forth by Lui et al. [66] and Madni et al. [68] center on the ability of DT to progress over time through the acquisition of updated information from the physical asset, which is then utilized to monitor the performance of the said asset. The definitions above are highly comprehensive and pertinent to the construction sector, as they relate to applying DT for proactive maintenance and ongoing monitoring within this specific field. A DT offers reliable models based on the quantity of collected data and presents a real-time depiction of the condition of the tangible entity.

4.3 Key Features

The development of DT has led to the creation of new platforms and routes that allow the execution of functionalities and services in the simplest way possible. One of the most innovative digital platforms that makes digital/virtual twins of physical assets has been named DT [69]. The DT functions facilitate system data evaluation and tracking through the interaction of the physical and digital realms.[6] The virtual model can seamlessly interact with the physical entity using effective communication platforms and rapid updates. As per the source, IoT has been recognized as a viable communication platform facilitated by sensors, cloud computing, and big data analytics [66]. The primary mechanism that underlies the functionality of DT is the communication of information and data from one endpoint to the other. During the asset's entire life cycle, information is continuously exchanged between a tangible and intangible assets [1, 86]. Through the iterative adaptation to operational modifications utilizing information and online data acquisition, a DT can emulate a physical asset by offering future prognostications. By anticipating necessary maintenance actions for a physical asset, maintenance strategies can transition from a reactive approach to a predictive one.

The linkage or interaction between the physical and virtual products of a DT, which necessitates the use of analytics at every stage, holds significant

---

[6] https://www.gavstech.com/wp-content/uploads/2017/10/Digital Twin Concept.pdf.

importance. Madni and colleagues (2018) identified several features of Decision Trees, as outlined in their study [68].

- A DT represents the distinct instance, performance, maintenance, repair history, health status, and other attributes of a physical structure.
- Preventive maintenance schedules are established based on the historical data of the physical structure.
- It makes use of the virtual model, to track, comprehend, and forecast future performance and maintenance patterns for physical assets.
- Developers and facility managers can keep an eye on the system's performance and potentially make adjustments to bring it in line with expectations.
- Using the interconnectivity facilitated by digital threads, monitoring the various phases comprising the life cycle of tangible assets becomes feasible.
- Predictive analytics can be employed to forecast system performance in the future by refining assumptions based on data collected from the physical structure.
- By troubleshooting broken remote equipment, it may be possible to perform remote maintenance.
- Integrating data obtained from the Internet of Things and data derived from physical assets is leveraged to enhance and streamline services and operational processes.
- The duration of the physical system's existence can be represented through a discrete-time model that emulates operational and maintenance information.

4.4 System Architecture

A thorough understanding of the dynamics necessary for DT functionality is crucial in order to address the system architecture. Engineering data, such as CAD files, product specifications, geometry models, material properties, validation results, IoT sensor readings, and simulation results, all play a role in the dynamics that determine the DT system's architecture [79]. Selecting an appropriate model for creating a Decision Tree is contingent upon its intended application. Effective system implementation will significantly influence the physical infrastructure's design, construction, and operation over its entire lifespan. According to the literature, the three primary constituents of DT are the tangible system, the intangible system, and the interconnecting communication channel [40]. The constituents of DT are comprised of various technological advancements that facilitate its implementation, including but not limited to mobile devices, cyber-physical systems, sensors, and communication networks. Cyber-physical systems facilitate the smooth integration of virtual and physical models [3, 4, 5]. To replicate the actions of a tangible entity through a digital representation, and offer appropriate remedies when necessary, sensors are affixed to the human structure [22]. The information produced by the interconnected sensors is transmitted to the cloud through the IoT. Furthermore, it is synchronized with the virtual system to ensure that the latest data regarding the physical system is available.

DT leverages the benefits of the IoT to streamline the transmission of information from rudimentary sensory data to sophisticated insights [66]. AI, machine learning, and simulation prowess are all must-haves for this method. Real-time updation of the virtual model is enabled by processing data and information obtained from the attached sensors. Therefore, a fundamental requirement for this procedure is high proficiency in simulation, machine learning, and AI. The ability to update the virtual model in real time is facilitated by processing data and information collected by the attached sensors. To establish a seamless communication pathway between the two domains, the DT above attributes entail extracting information from the tangible infrastructure and harmonizing archival and current data [100].

Data (such as structures, external data, and blueprint data), reasoning (such as AI/ML models or unpredictable rules), key performance indicators (such as efficiency, emissions, NORI, and safety indicators), and context (such as occupant behavior, system/device behavior, and workflows) are the four fundamental components of a DT identified by Castaldini (29). These elements are crucial for the successful implementation of a DT. The four fundamental components play a decisive role in shaping the system architecture and functionality of a DT. Tao and colleagues [100] have expanded upon the four fundamental components by categorizing them into six distinct domains. These areas include data measurement, interaction with the virtual product, behavior simulation, behavior control, data analysis, data integration, and visualization; virtual products; and behavior simulation.

As per the findings of the 2022 Chartered Institute of Arbitrators (CIArb) Webinar report, the worldwide construction industry is projected to experience a growth of $8 trillion by the year 2030, primarily propelled by China, the United States, and India.[7] As per the Global Construction 2030 report findings, it is anticipated that China, the United States, and India will spearhead global growth and contribute to 57% of the overall expansion. The report further forecasts that the worldwide construction output will surge 85% and reach

---

[7] https://www.intellectsoft.net/blog/8-emerging-construction-technology-trends/

$15.5 trillion by 2030. Cutting-edge technologies and construction software will be utilized throughout all phases of the construction process, including project planning, on-site work, and complementary stages. Advanced technologies and construction software will be employed throughout the project's various stages, including project planning, on-site work, and complementary phases.

Virtual Reality (VR) and Augmented Reality (AR) have been major innovations in recent construction technology, offering visualizations. The incorporation of VR and AR is anticipated to significantly impact the projected $8 trillion global growth of the industry by 2030. The potential for AR/VR technology to be utilized in remote site inspections has been noted to significantly reduce construction costs by up to 90% when implemented in 2022. Additionally, it promotes teamwork, communication, and safety for AEC employees.

4.4.1 Drones

Over a decade ago, drones were first used in construction. Commencing 2023 and beyond, we can anticipate the emergence of more sophisticated drones with a focus on AI. Using drones with 3D LIDAR scans and real-time aerial imagery is revolutionising the construction industry.

4.4.2 Blockchain Technology

In 2023 and beyond, Blockchain Technology is expected to enhance cost management and facilitate efficient procurement strategies throughout the (AEC) industry. Incorporating a technological innovation (comprising interconnected data blocks forming a digital ledger that encompasses all transactional and milestone records), into the construction sector is a relatively recent development, having occurred within the last decade. The analogy can be drawn between a physical chain and a project's transactions, where each link represents a particular transaction, and the system inherently maintains equilibrium. Unlike other alternatives, this option is characterized by its security, decentralization, and flexibility, allowing it to adapt to projects of varying scales.

4.4.3 DT

The implementation of DT represents the latest technological advancement in the AEC sector, aimed at resolving operational management challenges. The process entails using simulation techniques to generate a prototype of a building. The DT trends and functionality domain pertains to intelligent multidimensional digital models. In the upcoming years, the number of buildings experiencing operational issues is expected to decrease. This can be attributed to the use of DTs, which can simulate, predict, and provide informed decisions based on environmental circumstances. Designers and technicians conduct a performance analysis that considers occupants' behavior through patterns and spatial considerations.

4.4.4 3D Laser Scanner

The DT technology is a recent development in the AEC industry, focusing on managing operational issues. Simulation is utilised as a means of constructing a building prototype. DT's development and utility depend on the use of sophisticated multi-dimensional digital models. The utilisation of Decision Trees is expected to result in a decrease in operational challenges encountered by buildings beyond the year 2023. This is due to their capacity to imitate, forecast, and aid in decision-making based on actual situations. Designers and technicians utilise patterns and spatial considerations to conduct performance analysis and factor in occupant behavior. The technology in question was exemplified through the utilisation of Aiguilles-Queyras Hospital Center,[8] a medical facility in the French Alps. The 3D laser scanner manufactured by FARO[9] comprehensively scans the entire pre-existing hospital structure, thereby generating a novel model to facilitate the renovation process.

4.4.5 4D Simulation

Using 4D simulations is a contemporary technological approach, that aims to minimize expenses and time associated with onsite and offsite projects. This is a significant consideration, given the construction industry's persistent pursuit of efficiency. The Bentley pipeline project[10] for the Haweswater Aqueduct utilized 4D simulation, which allowed Mott MacDonald to save 20 days.

4.4.6 3D Printing

While less prevalent than BIM, 3D printing represents one of the latest technological advancements in the construction industry. In 1995, a mechanism was introduced to generate three-dimensional structures from digital models. The utilization of 3D software to produce 3D models has gained significant popularity in 2023. The utilization of 3D printing technology by the engineers at Arup facilitated the fabrication of steel nodes for a structure

---

[8] http://www.ch-aiguilles.fr/

[9] https://www.linkedin.com/company/faro-technologies/

[10] https://www.bimplus.co.uk/how-4d-helped-mott-mac-bentley-save-20-days-major/

that is distinguished by its lightweight properties.[11] Furthermore, Win Sun Decoration Design Engineering,[12] a Shanghai-based corporation utilized 3D printing technology to administer a blend of rapidly solidifying cement and repurposed constituents in a spreadable format.

5. Structured DT System Design

As per the academic discourse, a city represents a complex network of interrelated social, economic, and physical components [67]. Information and Communication Technology infrastructure in urban areas, facilitates the interconnection of diverse asset groups and the derivation of insights from multiple data sets, as noted in reference [93]. Consequently, an urban local body (ULB) may be regarded as a valuable entity that amalgamates diverse constituent assets, comprising edifices, amenities, transportation systems, and populace. Thus, a city-level DT refers to an interactive digital model of a city that integrates all its sub-DTs. Figure 2 illustrates the hierarchical parent-child relationship among DTs across multiple levels. To ensure the confidentiality of data within each data terminal, higher-level DTs (such as those at the city level) engage in bidirectional communication with lower-level sub-DTs (such as those at the building level). This communication involves requesting necessary data, addressing stakeholder requests, and providing tailored services.

In this chapter, a hierarchical architecture is shown for both the city and the building levels. The following sections comprise the processes of obtaining data, transmitting it, creating digital models, integrating data and models, and implementing service layers.

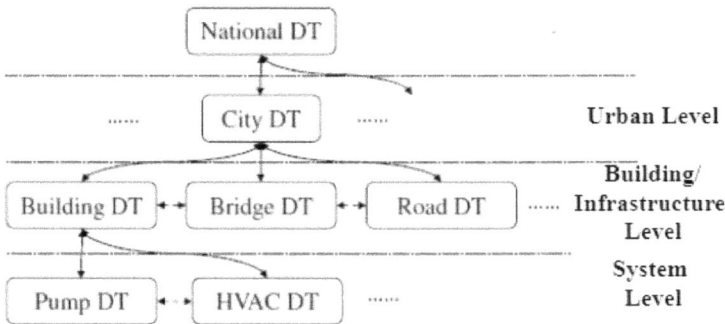

Figure 2. DT links and order between the different levels.

---

[11] https://www.arup.com/projects/3d-printed-concrete-house.
[12] https://www.designingbuildings.co.uk/wiki/3D printing in construction.

5.1 Data Acquisition Layer

Each DT is made up of a layer for getting data. Because there are so many kinds of data and so much of it at the city level, it is essential and hard to design a way to get data. This holds regardless of the data's nature, origin, structure, or content. Also, the subsets, like buildings and transportation systems, will have sub-DTs based on their daily operations roles. The sub-DTs can provide the city DT with requested data, information, or models. Various techniques are utilized for data collection, including but not limited to radio-frequency identification (RFID), image-based methods, distributed sensor systems, wireless communication, and mobile access (e.g., in Wi-Fi environments). Each twin uses the DT architecture based on different sub-DT levels, such as buildings, with real-time data collection, efficient data management, and integration [35, 51]. Figure 3 shows a close-up of the DT architecture used for buildings. Building DTs have the same architecture as city DTs, with a data acquisition layer, transmission layer, digital modeling layer, data/model integration layer (with a simulation engine and data analysis functions), and service layer (with space use and workplace design).

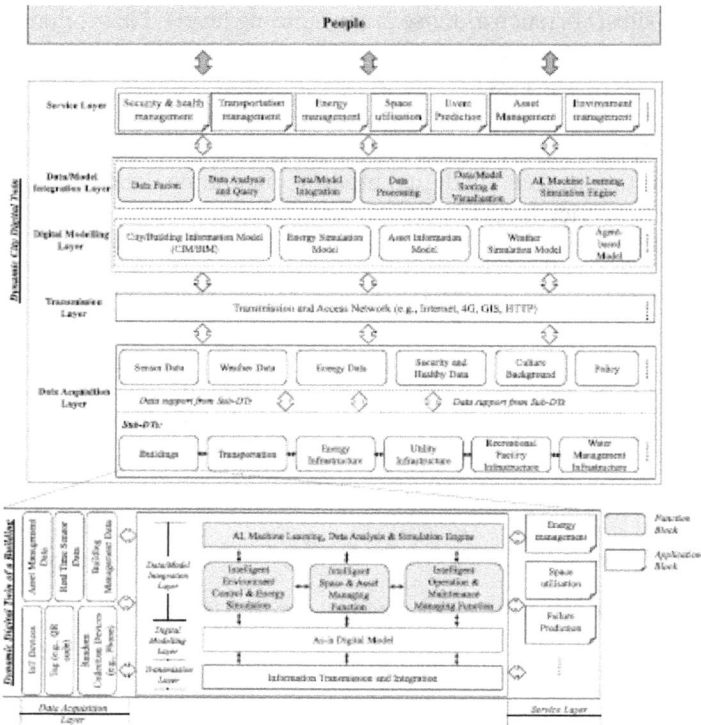

Figure 3. City and building DT system architecture.

5.2 Transmission Layer

The data collection and processing layers rely on the transmission layer to deliver the information to the next level for processing. In this layer, different kinds of communication technologies could be used, such as Wi-Fi, Zigbee, near-field communication (NFC), mobile-to-mobile (M2M), and Zwave, as well as 3G, 4G, long-term evolution (LTE), 5G, and low-power wide-area networks (LP-WAN) [37, 52]. Despite the rapid development of competing technologies, Wi-Fi remains the most common and widely adopted WLAN solution.

Despite being the most widely adopted technology, concerns regarding the security of Digital Technologies in developing urban areas utilizing Wi-Fi have rendered the unlicensed spectrum band problematic [42]. Light Fidelity (Li-Fi) and Low Power Wide Area Networks (LP-WAN) are two intriguing possibilities for developing distributed technologies (DTs) that require wide-area coverage both at the construction and city levels [58, 94]. This is particularly relevant when thinking about network efficiency and data transfer rate.

5.3 Digital Modeling Layer

The digital modeling layer comprises various digital representations of the physical asset, such as BIM and City Information Modeling (CIM), along with supplementary data that facilitates the upper layers, such as meteorological information and cultural contexts. BIM and CIM use similar ideas when discussing information models at the city level. They expand the ways that models, data, and methods like geographic information systems (GIS) can be used in city settings (like urban planning) to help people make decisions [38, 39]. In DTs, different models and types of models can be used for other goals. Real-time monitoring and control, an assets management model, a CIM for city and infrastructure planning, scenario modeling, and a decision support system are all excellent examples [21, 59]. To achieve unity with particular target applications, from singular infrastructures to complete edifices and municipalities, it is imperative to utilize a pre-established schema and systematic modeling procedures during the development of DT at construction and city scales [46]. The optimisation of city digital twins' performance relies heavily on the efficient and hierarchical design of models, data storage, integration, and query systems. The significance of this matter is heightened when confronted with intricate and extensive data sets, requiring the implementation of city-level storage and administration systems of considerable magnitude. Utilising cloud computing, storage, and data/model visualization can enhance data management at both the city and building levels, promoting a dynamic and practical approach.

5.4 Data/Model Integration Layer

The primary constituent of this structure is the stratum that amalgamates the data and model. This layer aims to integrate all available data sources within the designated data architecture. The stratum in question encompasses the requisite instruments for operations such as data processing, storage, analysis, integration, and manipulation and AI-facilitated knowledge acquisition that assist' decision-making [40]. As per the architectural framework, the actual state of building assets, including work orders, present maintenance data, and their respective statuses, alongside municipal assets, such as transportation conditions and energy usage, would be contemporaneously revised [64, 95].

5.5 Service Layer

The design is executed in the service layer, the uppermost tier of the DT architecture. To facilitate communication between people, communities, and the data/model integration layer, the system processes data from Knowledge Engines (KE). The service layer gives societal services and checks how well-built DTs work. This changes how happy people are with things like smart transportation, building sustainable communities, and taking care of the environment. Moreover, to enhance overall satisfaction, knowledge engineers (KEs) ought to integrate user feedback as an external source of knowledge.

6. Strengths of DT in Construction

In the real estate field, using DTs can facilitate a thorough assessment of the physical attributes of a property, enabling brokers or owners to gather and scrutinize relevant information about the asset's state and effectiveness. Therefore, the use of DTs in the construction and real estate industries has the potential to achieve previously un-imagined levels of efficiency in these areas (as well as time, expense, sustainability, and safety). As such, it is deemed a critical tool for the entire building sector.

6.1 Automate Project Control

Digital technologies can provide instantaneous feedback on the condition and efficacy of construction projects, thereby facilitating monitoring of their advancement. Sensor, drone, laser scanner, and monitoring system data can all be integrated into a virtual model [8]. Artificial Intelligence-enabled DT solutions can efficiently handle, evaluate, and exhibit the combined data in the form of an as-is-built model. Additionally, these solutions can provide hourly or daily comparisons to the baseline model. Monitoring and identifying any potential variances in budget or schedule during a construction project would aid project teams to devise and execute appropriate corrective measures, ultimately leading to the resolution of typical construction progress challenges.

6.2 Resource Planning and Logistics

The construction industry may generate significant waste due to the extra handling and transportation of supplies, tools, and labor. The implementation of DT solutions has the potential to enable a lean methodology in managing resources, thereby contributing to the reduction of wastage. The alternatives mentioned above can provide immediate supervision over-allocating resources and handling waste, thereby augmenting the efficiency and efficacy of the construction procedure. One instance of utilizing digital technologies in monitoring construction project advancement is the amalgamation of sensory information derived from construction equipment and vehicles. This approach enables the acquisition of instantaneous information regarding the location and utilization of assets, thereby facilitating the identification of possible obstructions or inadequacies. The benefits of decision trees is that they have the potential to assist the construction industry in mitigating resource depletion and improving time allocation, as evidenced by sources [6, 87, 109].

6.3 Construction Safety

The safety performance exhibited by the construction sector is below the optimal level. Integrating AI with cameras, sensors, and mobile devices through DTs technology provides a comprehensive safety framework for the construction sector.[13] DTs provide a real-time capability for site reconstruction, which the construction sector can utilise to facilitate the monitoring and tracking of the construction process. This characteristic can assist in detecting and resolving any discrepancies or variations from the established protocol that pose a safety risk [50]. This data makes it possible to avert the use of hazardous substances and involvement in risky undertakings within specified regions. Furthermore, the management can establish a mechanism for timely notifications that notifies them when an employee is at risk and transmits a signal to the employee's wearable gadget. Moreover, Decision Trees have the potential to detect hazardous actions and provide customized instruction to workers in simulated settings, obviating the necessity for costly and protracted on-site training and ultimately reducing the likelihood of mishaps [50].

---

[13] https://www.intellectsoft.net/blog/advanced-imaging-algorithms-for-digital-twin-reconstruction/

6.4 Quality Assurance

DT solutions can help improve quality assurance in construction by giving real-time information about how the different parts and systems used in the project are working and how they are behaving. DT technology employs image processing algorithms to assess the condition of different components and materials on a construction site by analyzing photographic or video imagery. Concrete condition assessment and crack detection in columns and other structures are among the potential applications of this technique. This can aid in identifying future issues or deficiencies and implementing corrective measures to ensure the quality of the construction procedure.

6.5 Assessing Building Performance

DT solutions can measure how well a building works, by simulating its behavior and performance under different conditions and scenarios. The objective of integrating data from various sources, including sensors, simulations, and BIM models, can be achieved through implementation. For example, DTs can facilitate the assessment of a building's energy efficiency by simulating energy generation and consumption. In addition, this approach enables the evaluation of the effects of various design and operational strategies on the building's energy performance. By optimising air, heat, and humidity circulation and evaluating the impacts of diverse design and control approaches, buildings can be assessed for their indoor environmental quality and comfort levels [50].

6.6 Anomaly Detection in Pumps

As part of the proposed DT architecture, the current service can detect spikes in vibration data that herald the onset of abnormal faults on Heating, Ventilation, and Air Conditioning (HVAC) pumps. To do this, we analyze vibration data that can be used as diagnostic indicators of the pumps' underlying mechanical health. Additional information about the shape and placement of buildings is available through BIM. In contrast, Building Management Systems (BMS) and live sensors primarily oversee the operational state of critical assets. The integration layer of data and model facilitates the astute extraction of data about pumps, which is made possible by implementing the Industry Foundation Classes (IFC) schema during the demonstration. The Cumulative Sum (CUSUM) charts, a prevalent method for detecting change points, are employed to examine the extracted pump data and see instances where the fundamental vibration symptom parameters diverge from their anticipated values.

6.7 Maintenance/Repair Prioritization

The allocation of resources is contingent to the prioritisation of maintenance tasks. As per Mobley's estimates [71], it can be inferred that a mere one-third of maintenance expenses are utilized efficiently. The present application uses a developed Decision Tree and leverages advancements in mobile communications, social networking, and machine learning to tackle the aforementioned municipal limitations. Furthermore, the provision of assets on the Internet is facilitated through the utilisation of asset tags and a digital asset profile. Individuals who utilise assets, can scan corresponding labels using a mobile application. This action enables viewing of digital profiles and the ability to append comments detailing any issues or problems encountered.[14] The proposed approach utilises a machine learning algorithm, operationalized by integrating data and model layers, to deduce the significance of individual asset defects. The feedback obtained from this process is subsequently employed as the input for the algorithm.

6.8 Green Urban Energy Planning

Urban energy planning has advanced beyond providing basic human needs and meeting societal expectations. It has now established a comprehensive approach to tackle ecological and energy-related concerns on an urban level, with the ultimate goal of achieving reduced carbon intensity. This particular application demonstrates the successful integration of sensing and computation capabilities within building DT. The integration is achieved through the utilisation of sequence-to-sequence long short-term memory (LSTM) technology, which approximates the characteristics of building energy demand patterns. By utilising this data, it is possible to determine the future requirements for the capacity and energy output of the energy supply infrastructure within the urban development framework. This objective's attainment can be facilitated by acquiring quantitative data on energy demand and the spatial distribution of the projected energy. Integrating an appropriate quantity of renewable energy sources can lead to an improved energy consumption pattern in urban areas.

7. Limitations of DT in Construction

Certain construction firms may encounter persistent challenges within the industry that must be surmounted before successfully implementing digital technologies. This section will address the primary challenges that require attention when implementing Decision Trees in the construction industry.

---

[14] https://itemit.com/

7.1 Lack of Contractual Awareness

Comprehending the intricacies of a contract constitutes a fundamental element of every construction undertaking, and the integration of novel technologies, such as DTs, can substantially alter the process. The utilization of Design Teams can serve as a means to ensure that contractual obligations about a building project are comprehended and adhered to by all parties involved. This is achieved by monitoring the project's scope and the client's primary objectives.

7.2 Lack of Data Understanding

Numerous construction firms need help utilising data efficiently due to inadequate tools for acquiring insights, disparate data management systems between information technology and operational technology, and limited access to real-time data or data within time constraints. One of the advantages of Matterport is its ability to integrate with various data streams seamlessly.

7.3 The Move to BIM

Numerous construction companies rely on 2D design technology for significant construction projects. Implementing a BIM model, connected to a virtual design and construction process necessitates using advanced 3D modeling technology and acquiring novel skill sets within an organization.

7.4 No Structure or Defined SOPs

The advent of DTs poses an additional obstacle for stakeholders, as specific construction projects may need to possess established protocols for data exchange and issue resolution. Decision Trees guarantee uniformity among all stakeholders involved in a project's lifespan, including contractors, inspectors, and clients, by establishing a unified and reliable information source.

8. Data Management Perspective on DT Development

The success of a DT at the city and building levels is contingent upon resolving four significant data management issues, as identified by the pilot. This is because a DT is fundamentally constructed upon data.

8.1 Data Integration

The process of identifying a data trustee presents several challenges in the realm of data management, particularly in the consolidation of data from disparate sources that are geographically dispersed and exhibit diversity. Using the provided DT, we can see how different data sources, such as real-time sensors, BMS, Cloud Services, and AMS, can be integrated to create a comprehensive picture. Various technological solutions are available to enable data integration from a technical standpoint. The technologies in question encompass a range of mechanisms, including extract, transform, and load (ETL), which facilitate data transfer between systems and service-oriented architectures that can expose data as a service [25]. Data virtualization, warehouses, and lakes [19] can also aid integration. However, in the context of organisational integration, employing a blend of various technologies as the only solution is often insufficient in addressing all issues [12]. Therefore, the primary obstacle in the development of DT lies in integrating diverse data sources and establishing connections with supplementary assets.

The utilization of big data is of paramount importance in digital transformations, characterized by their significant volume, velocity, and variety. Many DT functions would be unfeasible to execute without big data. The Semantic ETL workflow [16] represents a viable solution for data integration in digital transformation. The methodology above can be investigated to assimilate substantial amounts of data from various origins into a unified data structure, enabling quick data retrieval and bolstering innovative applications. During the transform phase of the ETL (Extract, Transform, Load) process, semantic technologies are utilised to establish a semantic data model and generate semantically linked data in the format of Resource Description Framework (RDF) triples. Subsequently, the triples above are retained within a repository of organized data, commonly called a data warehouse. The conventional approach is commonly utilised during the extraction and loading phases of the ETL process.

8.2 Diverse Source Data Systems

The fundamental data required for tracking and prediction algorithms are stored in source systems, typically distributed systems operating on various software platforms and database systems. The optimal performance of queries to extract data from these systems presents a formidable obstacle. Implementing a NoSQL engine within DynamoDB is appropriate for executing general parallel data queries, amidst a substantial influx of concurrent requests and storing data vastly. The significance of current and punctual data is particularly crucial within the framework of DTs. In certain circumstances, acquiring a consistent stream of real-time data, such as telemetry data, may be imperative under extraordinary conditions. Furthermore, urban downtown areas must retrieve data from their subordinate downtown areas.

The structure of data models often varies across systems due to the diverse storage options available to database designers for the same data type.

The manifestation of data is represented through different database tables, individual records, and the selection of specific attributes. Determining the originality of a data record about a specific machine in one system compared to another in a different approach presents a significant obstacle in the absence of a universally accepted standard or a globally unique identifier (GUID) for data records across diverse data sources. As a result, several terminologies, including entity linking, record linkage, entity resolution, data matching, and data duplication, are utilized in diverse systems [98].

An additional challenge involves the task of reconciling discrepancies in both the meaning and structure of the information. As an illustration, the classification of external pipework as part of a boiler may be encompassed by one reference while being excluded by another. The field of Master Data Management (MDM) provides solutions to the issues above by guiding on establishing consensus on data definitions and regulating its evolution and modification over time [76, 77]. Thus, to maintain uniformity in the terminology, it becomes imperative to harmonize the discrepancies in distinct parameters within databases. For instance, diverse measurement systems may be utilized, such as Celsius or Fahrenheit.

8.3 Data Synchronization

The data synchronisation process is not explicitly referenced within the preceding section of the demonstrator. The integration of data and models in the respective layer was primarily carried out through manual offline means. The synchronisation of multiple data copies is a crucial aspect of implementing a practical DT, as it determines the optimal timing and frequency required to provide the decision-makers with the most up-to-date information. The matter at hand is intricate, as it involves a trade-off between synchronisation expenses and the quality of data, which may suffer from staleness [82]. The costs incurred from utilizing resources such as computing facilities and personnel in the ITechnology field are encompassed within the ambit of synchronisation costs. The act of locking [104] systems to access data and the potential reduction in computing power can result in a considerable disruption cost to businesses. This is because any decrease in computer power can limit the energy available for essential business operations [82]. These factors highlight the potential impact of computing resources on business operations.

8.4 Data Quality

As per the assertion made by Wang and Strong [103], the concept of data quality can be defined as "fitness for use," which considers the data's intended use and adherence to the specified requirements. The utilisation of data in Decision Trees must support the concurrent facilitation of various applications, including but not limited to service decision-making and predictive analysis. Therefore, the information stored in a Decision Tree has the potential to degrade, rendering it unsuitable for use due to various factors.

Utilising this approach for acquiring data from source systems may result in a potential compromise of data quality. If the query for data extraction is formulated erroneously, inaccurate records may be procured. During semantic processing, incorrect data transformations may result in data loss. In the event of a flawless data extraction process, any errors in the source systems› data will still be transmitted to the data warehouse. Specific errors, such as invalid or improperly formatted data, can be identified and rectified during the transformation process [105]. Data technicians can utilize publicly available online data at the municipal level, such as meteorological predictions. The utilisation of online data may necessitate a distinct comprehension of data quality compared to conventional database systems, as the former may possess questionable quality [67].

9. BIM vs DT

BIM and DT technology are designed to enhance process transparency, cooperation, and efficiency. Nevertheless, these two technologies exhibit notable distinctions.[15]

---

[15] https://www.indovance.com/knowledge-center/building-information-modeling-bim-vs-digital-twins/

9.1 Definition

- **BIM** - BIM is a methodology that entails generating and managing digital representations of the aesthetic and functional characteristics of locations.
- **DT** - A digital representation of physical assets, processes, and systems, as well as their dynamics, is what is meant when someone refers to something as having a "Digital Twin."

9.2 Function

- **BIM** - During the design and construction phases of a project, BIM is intended to serve not only as a digital database but also as a tool that makes collaboration and visualization more accessible.
- **DT** - The purpose of DTs is to assist with the management and upkeep of the building.

10. Case Study

The Shanghai East Hospital, also known as the East Hospital, is a highly regarded general hospital in the Lujiazui commercial district of Shanghai, situated in the eastern region of China [81]. The rating received is of the highest academic caliber, an A-grade. Tongji University is the affiliated institution of the hospital.

The recently erected Case Building is a comprehensive hub for medical care, education, scientific inquiry, emergency response, preventative measures, and healthcare services. Unlike previous clinical facilities, the new tower emphasizes necessary public medical safety measures and advanced medical care services. The building consists of 26 levels, with 24 levels situated above ground and two underground. It provides sleeping arrangements for 500 individuals and covers a significant built-up area of around 83000 square meters [81].

Before the erection of the novel clinical tower, the mechanical and electrical apparatus at East Hospital consisted of complex operating systems that functioned autonomously. These systems comprise medical procedures, including gas, sewage treatment, and medical equipment. The insufficient operational and maintenance efficacy at East Hospital can be attributed to the demand for increased efficiency among middle and upper-level management personnel.

Furthermore, the hospital infrastructure requires continuous operation 24 hours a day, seven days a week. As a result, the department responsible for O&M management will encounter considerable pressure. Furthermore, traditional approaches to manual bookkeeping administration have become obsolete and non-intuitive, owing to the challenge of overseeing and scrutinising voluminous datasets from diverse origins.

The case study utilized the "continuous lifecycle integration method" as its main technical innovation. The approach entails the execution of a comprehensive timetable that enables the systematic and punctual incorporation of pertinent material across the complete duration. Typically, construction endeavors encompass a range of activities, including AEC, as well as Operation and Maintenance (O&M) tasks, which are integrated into their timeline. Thus, the complete life cycle may be partitioned into various stages, encompassing, though not restricted to, the design, construction, and operation and maintenance phases.

The Operations and Maintenance phase is a crucial stage in a building's lifecycle, given its substantial allocation of budget and lifetime costs. The domains of O&M and AEC are intrinsically linked. In the design phase, basic static information is determined, encompassing geometry, facility-related data, spatial partitioning, and electrical system logic, among other factors. During construction, intelligent hardware devices are installed or maintained to gather dynamic operational data.

Figure 4 illustrates the fundamental steps in developing a Decision Tree for the hospital framework. The procedure begins with a physical healthcare establishment and concludes with creating a comprehensive decision-making framework. The first phase of the process entailed the conversion of the hospital into a digital format through the utilisation of geometry modeling techniques. The geometry modeling process mainly entailed utilizing CAD drawings in the design phase. The utilization of the 3D point cloud method, which involves laser scanning or photogrammetry, has been explored in cases that involve complex mechanical areas. This was recorded and cited in the reference identified as [96]. Incorporating the BIM modeling process was a crucial milestone. As contact [102] indicates, it is commonly imperative to segregate dynamically generated information from static ontology data to uphold inherent data security and employ suitable storage strategies.

Consequently, an initial BIM was constructed by incorporating stationary models and ontology data. Using business systems and sensor devices, a dynamic BIM or digital shadow has been generated through process monitoring. The integration of analysis engines equipped with preexisting knowledge and AI models should be implemented immediately following the monitoring of incoming data. The on-site investigation played a crucial role in establishing a connection between the object mapping process and the actual hospital environment. It is imperative that accurate object mapping is utilized to transmit functional modules and diagnostic recommendations to a physical hospital facility.

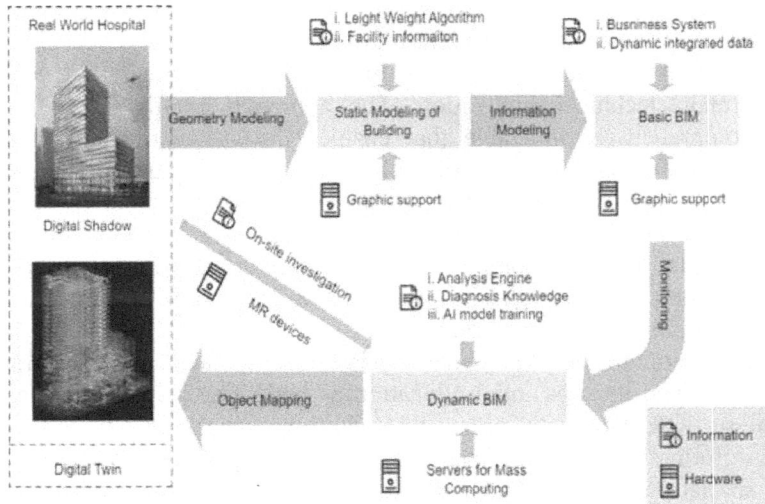

Figure 4. Main steps from reality to DT.

The fundamental concept underlying the design of the DT schema was the strategic execution of appropriate actions at the appropriate moments. From the onset of the design phase, it is imperative to integrate all contents, including static data, dynamic information, and potential diagnosis models from relevant stakeholders, in a timely manner throughout the various stages of the lifecycle.

11. Discussion

The DT has emerged as a powerful tool that has gained considerable traction across various industry sectors, garnering attention from scholars and corporations both domestically and internationally. The term above pertains to the electronic representation of a physical asset and its diverse instances, employing current information. At the outset, digital transformation requires the creation of virtual models of tangible assets, which are subsequently linked to their physical counterparts through sensors, intelligent devices (e.g., controllers and actuators), digital platforms, and communication networks. The sensors gather data from the tangible constituents and convey it to the digital medium via networks. The effectiveness and functional abilities of DT models rely on the amount and caliber of data and information accumulated over time regarding the physical asset. As per Buddoo's statement [84], automating the decision-making process for the DT of the physical system requires a pragmatic analysis of the collected data and information. The capability to provide real-time updates and virtual control of physical components is exclusively determined by data analytics. Unfortunately, utilising DT functionalities is more prevalent in industries other than the construction sector.

DT's adaptability is considered a critical factor in its widespread applicability and progress across diverse industries. According to Tao et al. [99], DT is currently providing benefits to the manufacturing sector, such as closed-loop manufacturing, optimization of product design, and intelligent maintenance, repair, and overhaul (MRO) processes. The manufacturing industry has adopted DT technology in four primary domains: DT-based product design, intelligent manufacturing in DT-enabled factories, product DT for usage monitoring, and DT as a facilitator for smart maintenance, repair, and overhaul (MRO) operations. The integration of DT applications within the manufacturing industry enables the identification and resolution of production bottlenecks while ensuring that final products meet user specifications with minimal resource and time wastage.

The construction sector has transitioned from traditional building design and construction techniques to integrating mechanical structures, commonly called smart buildings, which are furnished with many intelligent devices. The incorporation of smart devices aims to augment buildings' operational efficacy and ensure maximum user satisfaction. Numerous attempts have been made to implement DT technology in the construction sector. These efforts encompass diverse applications, such as monitoring the progress of construction activities [23] and generating a digital bridge model using a geometrical framework for DT implementation [24]. These software programs present potential benefits for enhancing DT applications in the construction sector and gaining a deeper understanding of their importance.

Digital technology presents potential advantages in the construction sector by enabling the assessment of a design's viability before its implementation. According to Siemens,[16] the early verification and validation of initial design choices is a crucial aspect of planning and coordinating multi-disciplinary solutions to enhance building performance. This process is considered an indispensable tool in achieving this objective. The potential advantages of utilising Design Thinking during the design phase are extensive and can have a positive impact throughout the entire lifecycle of a project or product. Real-time monitoring of construction projects enables the evaluation of compliance with established design standards. DT solutions have the potential to be implemented in post-construction scenarios to monitor building performance and facilitate virtual bi-directional coordination.

---

[16] https://www.siemens.com/global/en/products/buildings/digital-building-lifecycle/bim.html

12. Conclusion

The scientific concept of "Digital Twin" (DT) provides significant advantages and flexibility to physical objects/systems and their digital counterparts within the Industry 4.0 environment. Digital twin technologies in the construction industry enable the accumulation of data about an asset and its constituent parts, thereby facilitating the acquisition of pertinent insights into the support and its components over some time. This technology can reduce the monitoring and regulation of assets, as well as the optimisation of processes, throughout their operational lifespan. DT technology plays a crucial role in constructing facilities by predicting a facility's life cycle performance and generating economic value by optimizing potential future scenarios. Predicting the lifecycle enables adopting a Condition-Based Maintenance (CBM) program, which utilises up-to-date information to effectively allocate and enhance maintenance resources. Despite the widespread adoption of DT in various industries, such as construction, its full potential still needs to be realized. The present study offers a valuable overview for scholars and professionals regarding the current status of Digital Transformation across diverse sectors, with a particular emphasis on the construction industry.

The construction industry has been able to adopt DT more easily through cutting-edge technologies like BIM, point cloud segmentation, augmented reality, artificial intelligence, machine learning, data analytics, and sensors. Given the complexities of the construction industry, it is essential to implement DT applications to address current issues. Before beginning construction, conducting a design feasibility analysis with the help of DT technology can save time and money. When incorporated during the planning phase, it allows for tracking progress and identifying deviances from predetermined plans. To keep tabs on how well a building's many systems and appliances function, DT must be put into place wherever operations and maintenance are conducted. By providing real-time access to digital models of facilities, facilities managers can improve decision-making and the building's performance through bidirectional coordination based on up-to-date information. Thus, DT can be helpful when planning for the construction as well as during the execution, maintenance and monitoring of the structures.

References

[1] Abramovici, M., Gobel, J.C. and Dang, H.B. 2016. Semantic data management for the development and continuous reconfiguration of smart products and systems. CIRP Annals 65(1): 185–188.

[2] Agapaki, E., Miatt, G. and Brilakis, I. 2018. Prioritizing object types for modelling existing industrial facilities. Automation in Construction 96: 211–223.

[3] Akanmu, A., Anumba, C. and Messner, J. 2013. Scenarios for cyber-physical systems integration in construction. Journal of Information Technology in Construction (ITcon) 18(12): 240–260.

[4] Akanmu, A. and Anumba, C.J. 2015. Cyber-physical systems integration of building information models and the physical construction. Engineering, Construction and Architectural Management.

[5] Akanmu, A.A., Anumba, C.J. and Messner, J.I. 2012. An rtls-based approach to cyber-physical systems integration in design and construction. International Journal of Distributed Sensor Networks 8(12): 596845.

[6] Akanmu, A.A., Anumba, C.J. and Ogunseiju, O.O. 2021. Towards next generation cyber-physical systems and digital twins for construction. J. Inf. Technol. Constr. 26(Jul): 505–525.

[7] Alaka, H., Oyedele, L., Owolabi, H., Akinade, O., Bilal, M. and Ajayi, S. 2018. A big data analytics approach for construction firms failure prediction models. IEEE Transactions on Engineering Management 66(4): 689–698.

[8] Alizadehsalehi, S. and Yitmen, I. 2023. Digital twin-based progress monitoring management model through reality capture to extended reality technologies (drx). Smart and Sustainable Built Environment 12(1): 200–236.

[9] Janabi, S.A., Salman, M.A. and Mohammad, M. 2019. Multilevel network construction based on intelligent big data analysis. pp. 102–118. *In*: Big Data and Smart Digital Environment. Springer.

[10] Anumba, C.J., Akanmu, A. and Messner, J. 2010. Towards a cyberphysical systems approach to construction. pp. 528–537. *In*: Construction Research Congress 2010: Innovation for Reshaping Construction Practice.

[11] Anumba, C.J., Akanmu, A., Yuan, X. and Kan, C. 2021. Cyber—physical systems development for construction applications. Frontiers of Engineering Management 8: 72–87.

[12] Ara´ujo, T.B., Cappiello, C., Kozievitch, N.P., Mestre, D.G., Pires, C.E.S. and Vitali, M. 2017. Towards reliable data analyses for smart cities. pp. 304–308. *In*: Proceedings of the 21st International Database Engineering & Applications Symposium.

[13] Arayici, Y., Kiviniemi, A.O., Coates, S.P., Koskela, L.J., Kagioglou, M., Usher, C., O'Reilly, K. et al. 2011. Bim implementation and adoption process for an architectural practice.

[14] Azhar, S., Khalfan, M. and Maqsood, T. 2015. Bim: Now and beyond, construction economics and building. J. 12(4): 15–28.

[15] Bahmani, R. and Ouarda, T.B.M.J. 2021. Groundwater level modeling with hybrid artificial intelligence techniques. Journal of Hydrology 595: 125659.

[16] Bansal, S.K. and Kagemann, S. 2015. Integrating big data: A semantic extract-transform-load framework. Computer 48(3): 42–50.

[17] Barlish, K. and Sullivan, K. 2012. How to measure the benefits of bim—a case study approach. Automation in Construction 24: 149–159.

[18] Beveridge, S. 2012. Best Practices Using Building Information Modeling in Commercial Construction. Brigham Young University.

[19] Beyer, M., Thoo, E., Selvage, M. and Zaidi, E. 2017. Gartner research report: Magic quadrant for data integration tools. Gartner Research Rep (G00314940).

[20] Bilal, M., Oyedele, L.O., Qadir, J., Munir, K., Ajayi, S.O., Akinade, O.O., Owolabi, H.A., Alaka, H.A. and Pasha, M. 2016. Big data in the construction industry: A review of present status, opportunities, and future trends. Advanced Engineering Informatics 30(3): 500–521.

[21] Bolton, A., Enzer, M., Schooling, J. et al. 2018. The gemini principles: Guiding values for the national digital twin and information management framework. Centre for Digital Built Britain and Digital Framework Task Group.

[22] Boschert, S., Heinrich, C. and Rosen, R. 2018. Next generation digital twin. pp. 7–11. *In*: Proc. tmce, Las Palmas de Gran Canaria, Spain.

[23] Braun, A., Tuttas, S., Stilla, U. and Borrmann, A. 2018. Bim-based progress monitoring. pp. 463–476. Building Information Modeling: Technology Foundations and Industry Practice.
92 Blockchain and Digital Twin Enabled IoT Networks: Privacy and Security Perspectives

[24] Brilakis, I., Pan, Y., Borrmann, A., Mayer, H.G., Rhein, R., Vos, C., Pettinato, E. and Wagner, S. 2019. Built environment digital twining. International Workshop on Built Environment Digital Twinning presented by . . .. 2019.

[25] Budgen, D., Rigby, M., Brereton, P. and Turner, M. 2007. A data integration broker for healthcare systems. Computer 40(4): 34–41.

[26] Bughin, J., Kretschmer, T. and van Zeebroeck, N. 2021. Digital technology adoption drives strategic renewal for successful digital transformation. IEEE Engineering Management Review 49(3): 103–108.

[27] Cai, F., Ji, J.M., Jiang, Z.Q., Mu, Z.R., Wu, X., Zheng, W.J., Zhou, W.X., Tu, S.T. and Qian, X. 2018. Engineering fronts in 2018. Engineering 4(6): 748–753.

[28] Carls´en, A. and Elfstrand, O. 2018. Augmented construction: Developing a framework for implementing building information modeling through augmented reality at construction sites.

[29] Castaldini, F. 2019. How digital twin technology is central to smart buildings. Facility Executive.

[30] Chauhan, S., Patel, P., Delicato, F.C. and Chaudhary, S. 2016. A development framework for programming cyber-physical systems. pp. 47–53. *In*: 2016 IEEE/ACM 2nd International Workshop on Software Engineering for Smart Cyber-Physical Systems (SEsCPS).

[31] Chen, X., Shi, Q., Yang, L. and Xu, J. 2018. ThriftyEdge: Resource-efficient edge computing for intelligent IoT applications. IEEE Network 32(1): 61–65.

[32] Cheng, J., Qi, Q., Zhang, M. and Tao, F. 2018. Digital twin-driven product design, manufacturing and service. e International Journal of Advanced Manufacturing Technology 94: 3563–3576.

[33] Du, X., Xu, H. and Zhu, F. 2021. A data mining method for structure design with uncertainty in design variables. Computers & Structures 244: 106457.

[34] Saddik, A.E. 2018. Digital twins: The convergence of multimedia technologies. IEEE Multimedia 25(2): 87–92.

[35] Fattah, S.M.M., Sung, N.M., Ahn, I.Y., Ryu, M. and Yun, J. 2017. Building iot services for aging in place using standard-based iot platforms and heterogeneous iot products. Sensors 17(10): 2311.

[36] Formoso, C.T., Tzortzopoulos, P. and Forgues, D. 2020. Lean and bim meet social sciences: new perspectives in construction engineering and management. Canadian Journal of Civil Engineering 47(2): v–vi.

[37] Ge, X., Tu, S., Mao, G., Wang, C.X. and Han, T. 2016. 5g ultra-dense cellular networks. IEEE Wireless Communications 23(1): 72–79.

[38] Gil, J., Almeida, J. and Duarte, J.P. 2011. The backbone of a city information model (cim). Respecting Fragile Places: Education in Computer Aided Architectural Design in Europe, pp. 143–151.

[39] Gil, J., Beirao, J., Montenegro, N. and Duarte, J. 2010. Assessing computational tools for urban design: towards a "city information model". pp. 316–324. In 28th Conference on Future Cities.

[40] Glaessgen, E. and Stargel, D. 2012. The digital twin paradigm for future nasa and us air force vehicles. In 53rd AIAA/ASME/ASCE/AHS/ASC structures, structural dynamics and materials conference 20th AIAA/ASME/AHS adaptive structures conference 14th AIAA, page 1818.

[41] Grieves, M. 2015. Digital twin: Manufacturing excellence through virtual factory replication, michael w. GRIEVES, LLC, Cocoa Beach, Florida, USA.

[42] Grieves, M. 2014. Digital twin: manufacturing excellence through virtual factory replication. White Paper 1(2014): 1–7.

[43] Grieves, M. and Vickers, J. 2017. Digital twin: Mitigating unpredictable, undesirable emergent behavior in complex systems. Transdisciplinary Perspectives on Complex Systems: New Findings and Approaches, pp. 85–113.

[44] Grieves, M.W. 2005. Product lifecycle management: the new paradigm for enterprises. International Journal of Product Development 2(1-2): 71–84.

[45] Guo, Y., Xu, J., Qiu, S., Hu, Y., Zhang, X., Liang, Y., Gao, H. and Zhang, Y. 2021. Research on informatization construction technology based on computer statistics. In IOP Conference Series: Earth and Environmental Science, volume 632, page 042058. IOP Publishing.

[46] Habibi, M., Fazli, M.A. and Movaghar, A. 2019. Efficient distribution of requests in federated cloud computing environments utilizing statistical multiplexing. Future Generation Computer Systems 90: 451–460.

[47] Halttula, H., Haapasalo, H. and Silvola, R. 2020. Managing data flows in infrastructure projects-the lifecycle process model. Journal of Information Technology in Construction (ITcon) 25(12): 193–211.

[48] Hellyer, P. 2019. Digital methods as good as traditional construction. British Dental Journal 227(7): 585–585.

[49] Hodge, V.J., O'Keefe, S., Weeks, M. and Moulds, A. 2014. Wireless sensor networks for condition monitoring in the railway industry: A survey. IEEE Transactions on Intelligent Transportation Systems 16(3): 1088–1106.

[50] Hou, L., Wu, S., Zhang, G., Tan, Y. and Wang, X. 2020. Literature review of digital twins applications in construction workforce safety. Applied Sciences 11(1): 339.

[51] Hu, Z.Z., Tian, P.L., Li, S.W. and Zhang, J.P. 2018. Bimbased integrated delivery technologies for intelligent mep management in the operation and maintenance phase. Advances in Engineering Software 115: 1–16.

[52] Huang, J., Qian, F., Gerber, A., Mao, Z.M., Sen, S. and Spatscheck, O. 2012. A close examination of performance and power characteristics of 4g lte networks. pp. 225–238. In Proceedings of the 10th International Conference on Mobile Systems, Applications, and Services.

[53] Huang, M.Q., Nini´c, J. and Zhang, Q.B. 2021. Bim, machine learning and computer vision techniques in underground construction: Current status and future perspectives. Tunnelling and Underground Space Technology 108: 103677.

[54] CRC Construction Innovation. 2007. Adopting bim for facilities management: Solutions for managing the sydney opera house. Cooperative Research Center for Construction Innovation, Brisbane, Australia, Industry Publication.

[55] Javed, Y., Felemban, M., Shawly, T., Kobes, J. and Ghafoor, A. 2020. A partition-driven integrated security architecture for cyberphysical systems. Computer 53(3): 47–56.

[56] Ke, Y., Qiu, Y. and Yang, Z. 2021. Study on the maturity of construction technology in china. In IOP Conference Series: Earth and Environmental Science, 632: 022053. IOP Publishing.

[57] Khajavi, S.K., Motlagh, N.H., Jaribion, A., Werner, L.C. and Holmstrom, J. 2019. Digital twin: vision, benefits, boundaries, and creation for buildings. IEEE Access 7: 147406–147419.

[58] Khandal, D. and Jain, S. 2014. Li-fi (light fidelity): The future technology in wireless communication. International Journal of Information & Computation Technology 4(16): 1687–1694.

[59] Kim, K., Cho, Y.K. and Kim, K. 2018. Bim-based decision-making framework for scaffolding planning. Journal of Management in Engineering 34(6): 04018046.

[60] Klimova, E.V., Semeykin, A.Y., Sinebok, D.A. and Khomchenko, Y.V. 2020. The modern information technologies in construction for improving occupational safety. In IOP Conference Series: Materials Science and Engineering, 945: 012026. IOP Publishing.

[61] Kritzinger, W., Karner, M., Traar, G., Henjes, J. and Sihn, W. 2018. Digital twin in manufacturing: A categorical literature review and classification. Ifac-PapersOnline 51(11): 1016–1022.

[62] Kuzina, O. 2020. Information technology application in the construction project life cycle. In IOP Conference Series: Materials Science and Engineering 869: 062044. IOP Publishing.

[63] Lim, K.Y.H., Le, N.T., Agarwal, N. and Huynh, B.H. 2021. Digital twin architecture and development trends on manufacturing topologies. Implementing Industry 4.0: The Model Factory as the Key Enabler for the Future of Manufacturing, pp. 259–286. 94 Blockchain and Digital Twin Enabled IoT Networks: Privacy and Security Perspectives

[64] Lin, J., Zha, L. and Xu, Z. 2013. Consolidated cluster systems for data centers in the cloud age: a survey and analysis. Frontiers of Computer Science 7: 1–19.

[65] Lin, Y., Yang, Q. and Guan, G. 2019. Automatic design optimization of swath applying cfd and rsm model. Ocean Engineering 172: 146–154.

[66] Liu, Z., Meyendorf, N. and Mrad, N. 2018. The role of data fusion in predictive maintenance using digital twin. In AIP Conference Proceedings 1949: 020023. AIP Publishing LLC.

[67] Lukyanenko, R., Parsons, J. and Wiersma, Y.F. 2014. The iq of the crowd: Understanding and improving information quality in structured user-generated content. Information Systems Research 25(4): 669–689.

[68] Madni, A.M., Madni, C.C. and Lucero, S.D. 2019. Leveraging digital twin technology in model-based systems engineering. Systems 7(1): 7.

[69] Hernandez, V.C., Neely, A., Ouyang, A., Burstall, C. and Bisessar, D. 2019. Service business model innovation: the digital twin technology.

[70] Akanmu, A., Anumba, C. and Messner, J.I. 2011. Mechanisms for bi-directional coordination between virtual design and the physical construction.

[71] Mobley, R.K. 2002. An Introduction to Predictive Maintenance. Elsevier.

[72] Munoz, I., Madrid, J.A., Muniz, M.M., Uhart, M., Canou, J., Martin, C., Fabritius, M., Calvo, L., Poudelet, L., Cardona, R. et al. 2021. Life cycle assessment of integrated additive–subtractive concrete 3d printing. The International Journal of Advanced Manufacturing Technology 112: 2149–2159.

[73] Nassereddine, H., Veeramani, D. and Hanna, A. 2019. Augmented reality-enabled production strategy process. In ISARC. Proceedings of the International Symposium on Automation and Robotics in Construction 36: 297–305. IAARC Publications.

[74] Negri, E., Fumagalli, L. and Macchi, M. 2017. A review of the roles of digital twin in cps-based production systems. Procedia Manufacturing 11: 939–948.

[75] Nical, A.K. and Wodynski, W. 2016. Enhancing facility management through bim 6d. Procedia Engineering 164: 299–306.

[76] Otto, B. 2012. How to design the master data architecture: Findings from a case study at bosch. International Journal of Information Management 32(4): 337–346.

[77] Otto, B., Huner, K.M. and Osterle, H. 2012. Toward a functional reference model for master data quality management. Information Systems and e-Business Management 10: 395–425.

[78] Pachon, P., Castro, R., Macias, E.G., Compan, V. and Puertas, E. 2018. E. Torroja's bridge: Tailored experimental setup for shm of a historical bridge with a reduced number of sensors. Engineering Structures 162: 11–21.

[79] Padmanabhan, A. and Zhang, J. 2018. Cybersecurity risks and mitigation strategies in additive manufacturing. Progress in Additive Manufacturing 3: 87–93.

[80] Palit, S. and Datta, A. 2017. Emergence of digital twins. Journal of Innovation Management 5: 14–34.

[81] Peng, Y., Zhang, M., Yu, F., Xu, J. and Gao, S. 2020. Digital twin hospital buildings: an exemplary case study through continuous lifecycle integration. Advances in Civil Engineering 2020: 1–13.

[82] Qu, X.S. and Jiang, Z. 2018. A time-based dynamic synchronization policy for consolidated database systems. MIS Quarterly, Forthcoming.

[83] Rajaee, T., Ebrahimi, H. and Nourani, V. 2019. A review of the artificial intelligence methods in groundwater level modeling. Journal of Hydrology 572: 336–351.

[84] Madubuike, O.C., Chimay, J.A. and Khallaf, R. 2022. A review of digital twin applications in construction.

[85] Rivera, M.L., Vielma, J.C., Herrera, R.F., Carvallo, J. et al. 2019. Methodology for building information modeling (bim) implementation in structural engineering companies (secs). Advances in Civil Engineering.

[86] Rosen, R., Wichert, G.V., Lo, G. and Bettenhausenj, K.D. 2015. About the importance of autonomy and digital twins for the future of manufacturing. Ifac-Papersonline 48(3): 567–572.

[87] Sacks, R., Brilakis, I., Pikas, E., Xie, H.S. and Girolami, M. 2020. Construction with digital twin information systems. DataCentric Engineering 1: e14.

[88] Sacks, R., Kedar, A., Borrmann, A., Ma, L., Brilakis, I., Huthwohl, P., Daum, S., Kattel, U., Yosef, R., Liebich, T. et al. 2018. Seebridge as next generation bridge inspection: Overview, information delivery manual and model view definition. Automation in Construction 90: 134–145.

[89] Schroeder, G.N., Steinmetz, C., Pereira, C.E. and Espindola, D.B. 2016. Digital twin data modeling with automationml and a communication methodology for data exchange. IFAC-PapersOnLine 49(30): 12–17.

[90] Schweigkofler, A., Monizza, G.P., Domi, E., Popescu, A., Ratajczak, J., Marcher, C., Riedl, M. and Matt, D. 2018. Development of a digital platform based on the integration of augmented reality and bim for the management of information in construction processes. In Product Lifecycle Management to Support Industry 4.0: 15th IFIP WG 5.1 International Conference, PLM 2018, Turin, Italy, July 2–4, 2018, Proceedings 15, pages 46–55. Springer.

[91] Sfar, A.R., Challal, Y., Moyal, P. and Natalizio, E. 2019. A game theoretic approach for privacy preserving model in iot-based transportation. IEEE Transactions on Intelligent Transportation Systems 20(12): 4405–4414.

[92] Shafto, M., Conroy, M. and Doyle, R. 2010. Modeling, simulation, information technology & processing-tall.

[93] Silva, B.N., Khan, M. and Han, K. 2018. Towards sustainable smart cities: A review of trends, architectures, components, and open challenges in smart cities. Sustainable Cities and Society 38: 697–713.

[94] Silva, B.N., Khan, M. and Han, K. 2018. Towards sustainable smart cities: A review of trends, architectures, components, and open challenges in smart cities. Sustainable Cities and Society 38: 697–713.

[95] Silva, B.N., Khan, M. and Han, K. 2020. Integration of big data analytics embedded smart city architecture with restful web of things for efficient service provision and energy management. Future Generation Computer Systems 107: 975–987.

[96] Stojanovic, V., Trapp, M., Richter, R., Hagedorn, B. and Dollner, J. 2018. Towards the generation of digital twins for facility management based on 3d point clouds. Management 270: 279.

[97] Succar, B. 2009. Building information modelling framework: A research and delivery foundation for industry stakeholders. Automation in Construction 18(3): 357–375.

[98] Talburt, J.R. 2011. Entity Resolution and Information Quality. Elsevier.

[99] Tao, F., Cheng, J., Qi, Q., Zhang, M., Zhang, H. and Sui, F. 2018. Digital twin-driven product design, manufacturing and service with big data. The International Journal of Advanced Manufacturing Technology 94: 3563–3576.

[100] Tao, F., Zhang, H., Liu, A. and Nee, A.Y.C. 2018. Digital twin in industry: State-of-the-art. IEEE Transactions on Industrial Informatics 15(4): 2405–2415.

[101] Tsai, Y.C., Wu, Y.C. and Price, G. 2018. A cost-effective and objective full-depth patching identification method using 3d sensing technology with automated crack detection and classification. Transportation Research Record 2672(40): 50–58.

[102] Uhlemann, T.H.J., Lehmann, C. and Steinhilper, R. 2017. The digital twin: Realizing the cyber-physical production system for industry 4.0. Procedia Cirp 61: 335–340.

[103] Wang, R.Y. and Strong, D.M. 1996. Beyond accuracy: What data quality means to data consumers. Journal of Management Information Systems 12(4): 5–33.

[104] Woodall, P. 2017. The data repurposing challenge: new pressures from data analytics. Journal of Data and Information Quality (JDIQ) 8(3-4): 1–4.

[105] Woodall, P., Oberhofer, M. and Borek, A. 2014. A classification of data quality assessment and improvement methods. International Journal of Information Quality 16, 3(4): 298–321. 96 Blockchain and Digital Twin Enabled IoT Networks: Privacy and Security Perspectives

[106] Xu, L.D., Xu, E.L. and Li, L. 2018. Industry 4.0: state of the art and future trends. International Journal of Production Research 56(8): 2941–2962.

[107] Xu, X., Ma, L. and Ding, L. 2014. A framework for bim-enabled life-cycle information management of construction project. International Journal of Advanced Robotic Systems 11(8): 126.

[108] Yan, H., Yu, P. and Long, D. 2019. Study on deep unsupervised learning optimization algorithm based on cloud computing. pp. 679–681. *In*: 2019 International Conference on Intelligent Transportation, Big Data & Smart City (ICITBS). IEEE.

[109] Yang, J., Park, M.W., Vela, P.A. and Fard. M.G. 2015. Construction performance monitoring via still images, time-lapse photos, and video streams: Now, tomorrow, and the future. Advanced Engineering Informatics 29(2): 211–224.

[110] Yang, X.N., Wang, W., Xu, X.F., Pang, G.R. and Zhang, C.L. 2018. Research on the construction of a novel cyberspace security ecosystem. Engineering 4(1): 47–52.

[111] Yi, C.W., Chuang, Y.T. and Nian, C.S. 2015. Toward crowdsourcing-based road pavement monitoring by mobile sensing technologies. IEEE Transactions on Intelligent Transportation Systems 16(4): 1905–1917.

[112] Zhang, D.S. 2021. Scheme comparison and key construction technology for swivel rider cap of some super major bridge. In IOP Conference Series: Earth and Environmental Science 626: 012010. IOP Publishing.

[113] Zhao, X., Jia, Y., Li, A., Jiang, R. and Yichen Song. 2020. Multisource knowledge fusion: a survey. World Wide Web 23: 2567–2592.

[114] Zhao, Y.M., Han, Y., Kou, Y.Y., Li, L. and Du, J.H. 2021. Three-dimensional, real-time, and intelligent data acquisition of large deformation in deep tunnels. Advances in Civil Engineering 2021: 1–11.

[115] Zheng, P. and Lim, K.Y.H. 2020. Product family design and optimization: a digital twin enhanced approach. Procedia CIRP 93: 246–250.

[116] Zheng, Y., Yang, S. and Cheng, H. 2019. An application framework of digital twin and its case study. Journal of Ambient Intelligence and Humanized Computing 10: 1141–1153.

[117] Zhuang, C., Liu, J. and Xiong, H. 2018. Digital twin-based smart production management and control framework for the complex product assembly shop-floor. The International Journal of Advanced Manufacturing Technology 96: 1149–1163.

CHAPTER 5

Smart Agricultural Platform to Increase the Human Life Span Using Blockchain Technique

D. Menaka, S. Vigneshwari,* B. Sathiyaprasad and *S. Sreeji*

IIIIIIIIIIIIIIIIIIIIIIIIIIIIIIIIIIIIIIIIIIIIIIIIIIIIIIIIIIIIIIIIIIIIIIIIIIIIIIIIIIIIIIIIIIIIIIIIIIIIIIIIIIIIIIIIIIIIIIIIIIIIIIIIIIIIIIIIIIIIIIIIII

1. Introduction

Different economic areas of the country, including health care, farming, and business, were drawn to the advancement of technology in order to grow their enterprises and increase their profits. By utilising the sensing capabilities of IoT sensors, the advent of IoT in the agricultural industry provides a plethora of farming programs, like soil surveillance, crop value evaluation as well as water supply for cultivation. Similarly, population growth and environmental issues like shortages of water, changes in the climate, and warming oceans have an impact on crop output as a whole [1]. According to a study from the UN, the world's total population has increased by 2 billion people in the last 30 years. In 2050, it will rise from 7.7 billion to 9.7 billion, in accordance with their research [2]. As a result, there are now significant concerns about how to meet the rising food demand due to population growth. Hence, there is now a real issue regarding how to meet the rising need for food due to population growth.

Farmers must boost agricultural output by increasing farmland or by utilising the inorganic way of farming in order to meet the growing demand for food. Overusing herbicides and fertilisers, however, lowers agricultural excellence and poses [3, 4] a risk to humanity. Also, eating unhealthy food

Department of Computer Science and Engineering, Sathyabama Institute of Science and Technology, Chennai, Tamilnadu, India.
* Corresponding author: vigneshwari.cse@sathyabama.ac.in

produced by non-organic farming has negative effects on the body, including nausea, vertigo, nervousness, constipation, and nausea, which shortens life expectancy. As a result, maximum residual limits for every pesticide utilised for crops have been set by various food regulatory authorities, including "the Food Safety and Standards Authority of India (FSSAI)", and "Committee on the Environment, and Public Health and Food Safety" [5]. Farmers nevertheless continue to use less expensive, off-patent pesticides like "lindane and dichlorodiphenyltrichloroethane (DDT)", which are difficult to remove from the soil as well as from water [6].

Ingenious solutions kept numerous researchers from addressing the aforementioned problems with smart technology in agriculture. For instance, Hassaan et al. [7] investigated the environmental impacts and pesticide toxicity parameters. They researched a number of methods for estimating how few pesticides are present in crops. Additionally, they talked about how pesticides can affect human beings through a variety of routes, including inhalation, skin contact, and ingestion. Edge computing as well as blockchain technology has recently drawn attention in part thanks to providing possibilities for data protection and storage [8]. Taking into account this context, the scientists of [8] proposed an information discovery technique that incorporates a hyper-ledger, which offers effectual storage volume and maintains food confidentiality. Additionally, the authors suggested [9] a small signing approach for the network of blockchain connections employing edge computing as well as a hash algorithm to address data efficiency difficulties in detecting crops [10]. The proposed system is contrasted with well-known digital form signatures. The efficiency of the short-form signature technique, which reduces the data storage reaction time, is demonstrated by their findings.

This method inspired us to develop a smart farming system based on blockchain and artificial intelligence (AI) that effectively anticipates the abuse of insecticides in the crop as well as maximises the expectancy of human life. Initially, the collected crop records included environmental elements like hydrogen potential (pH) levels, temperature, and rainfall, agricultural characteristics including pesticides, pesticide threshold, crop and soil type. The dataset was then pre-processed by utilising normalisation approaches, missing values were filled in using centralised tendency measures as well as irrelevant features. The pre-processed dataset is further separated as testing and training datasets, with the training dataset being sent to several AI models for training, including Multilayer perceptron (MLP), support vector machines (SVM), Long and Short Term Memory (LSTM), perceptrons, and Logistic Regression (LR). On the other side, the accuracy of the above-mentioned technique is verified using the tested dataset. To avoid manipulation in data assaults when invaders change the pesticide threshold in order to confuse regulatory agencies, the projected data must be protected. Because of this, the proposed framework incorporates blockchain technology to store the forecasted data in the distributed blockchain ledger. Additionally, adopting 6G offers characteristics such as extremely low latency, high stability, ubiquitous high data rates as well as availability, which improves the performance of the proposed framework.

2. Related Works

Numerous studies have been conducted in the agricultural field that aid in cost calculation, weed detection, and crop and pesticide prediction. The researcher employs several methods and uses algorithms that revolutionise the agriculture industry all over the world. To improve the quality and volume of the crop, they use contemporary information and communication techniques like "machine learning (ML)", edge computing, "deep learning (DL)" as well as IoT. An improved agricultural production pest detection model, for instance, was presented by Jiao et al. [11]. They sought to address the low accuracy and efficiency problems in the conventional agriculture system. To do that, they used the pest dataset, which has 20 k photos as well as 24 classes, and the DL model. For the purpose of identifying agricultural pests, they presented the "anchor-free region convolutional neural network (AF-RCNN)" [12–13], "faster R-CNN", and "you only look once (YOLO)" models. They haven't, however, looked into how pests affect both the relevant crop and people. Then, for the detection of pests, Chen et al. [13] suggested "YOLOv3 and LSTM" [14–16]. They lessen crop loss and the harm that pesticide use causes to the environment. For crop recommendation, the author Priyadharshini et al. introduced the AI-based categorisation technique in [17]. They discovered the right soil characteristics, planting window, and geographic location to boost output and lower crop failure.

An AI-based technique that forecasts environmentally friendly crops based on the fertility of the soil was integrated by Nalwanga et al. [18]. They combined hardware "STM32F401CC and soil sensors" with software "STM32CubeIDE to monitor" the minerals in the soil while using the least amount of fertiliser. Then, Ref. [19] put forth a "crop yield prediction model based on artificial neural networks (ANN)". They used pH sensors as well as moisture sensors to gather their findings. Their suggested technique discusses the sensor's interfaced ANN network and produces effective outcomes. DL-based strategies were used by Jin et al. [20] to find the weeds in the crop. They suggested the CenterNet technique, which can recognise weeds and detect vegetables in a vegetable plantation. They also suggested SVM, Naive Bayes as well as KNN, classification algorithms in [21] along with a GPS and IoT module to prophecy soil quality as well as identify the kind and quantity of fertiliser and pesticides used in a crop. For the classification and detection of insects, Kasinathan et al. [22] introduced SVM, ANN, Naive Bayes, CNN as well as K-nearest neighbour (KNN), They examined the various classifiers and found CNN to be the most accurate and best-suited model.

KNN, random forest as well and SVM classifiers have recently been proposed by Sreedevi et al. [23] for the prediction of plant and crop diseases. The farmer can rent from them a tractor, a roller, a seed drill, a watering system as well as a harvester. Prior to the cultivation of crops, they suggest the crop grown, forecast plant diseases, and display the soil testing facilities. Additionally, Elakkiya et al. [24] reported AI-based methods for determining soil quality and disease-borne pathogens. In certain research studies, the authors place more emphasis on crop recommendations, weed identification, and insect and weed identification than they do on pesticides and their application in crops. Due to this, Durai et al. [25] proposed employing AI-based techniques to recommend crops and herbicides, and to identify weeds, and estimate costs. The use of CNN for pest classification was suggested by Setiawan et al. [26]. They categorise insect species to decide on the best preventive measures and lessen agricultural damage. The aforementioned research, however, did not take into account how pesticides affect both crops and people.

3. Methodology with Problem Formulation

Here, an outline of the methodology that predicts the optimal levels of pesticides to preserve human life expectancy. In the suggested system, we took into account the farm as $f \in f_1, f_2...f_i$ and the produce as $Crp \in Crp_1, Crp_2... Crp_n$. Here, farmer f_1 cultivates a variety of crops, including Crp_1, Crp_2, and Crp_3. However, the crop quality of Crp_1 is harmed by various insects $i_1, i_2...i_a$ and weeds $w_1, w_2...w_b$. The following diagram represents the aforementioned entity.

$$i_1, w_j \xrightarrow{\quad damage \quad} Crp_1 \tag{1}$$

$$\forall i = 1,2,....,a \& j = 1,2,....,b \tag{2}$$

To decrease crop damage from rodents, insects as well as weeds, and increase the yield of vegetables, fruits, as well as other products with quality, we employ a variety of pesticides, such as $p \in p_1, p_2...p_m$. However, excessive pesticide use has a negative impact on both human health H_h and the environment $Envt$.

$$fr_1 \xrightarrow{\quad p_1 \cdot p_2 p_3 \quad} Crp_1 \tag{3}$$

$$p_1, p_2, p_3 \xrightarrow{\quad harm \quad} h, Envt \tag{4}$$

Every pesticide has a predetermined threshold number known as p_{thr}, which is the maximum amount of pesticide that can be used on a crop. To maximise human life expectancy, we used AI algorithms to forecast the pesticide threshold in crop output. As a criterion for human life expectancy, we took into account the pesticide threshold for safe human consumption as well. The following are the effects of pesticide use on human life expectancy.

$$For, p < p_{thr} \rightarrow \omega = high \tag{5}$$

$$p > p_{thr} \rightarrow \omega = low \tag{6}$$

where p is the number of pesticides (m) that were applied to the specific crop. p_{thr} denotes the applicable pesticide's threshold value, and designates the expectancy of the human life parameter. The crop value is sent to the blockchain layer along with the relevant pesticides in this instance.

On a frontier server that offers minimal computing complexity when analysing information training, the suggested system handles information collecting, pre-processing, splitting, as well as validation. After gathering the crop information, we used pre-processing methods such normalisation as well as outlier and duplicate removal. Users must manage vast amounts of data and carry out data analytics for pesticide as well as crop forecast as well as cost of production assessment in the agriculture sector. They need the data to be authenticated and safe for that. To ensure security, reliability, and dependability in smart agriculture, we suggested using a blockchain system based on IPFS. Additionally, it faces data tampering attempts that change the pesticide threshold value to evade food regulatory agencies. In order to analyse the human life expectancy depending on the appropriate pesticide p used on crops *Crp*, the system model illustrates the detrimental effects of pesticides on the human body. Pesticide use has a negative impact on the human body when it reaches the threshold limit, as was stated in Eq. (6). As a result, the following is an explanation of the human life expectancy problem.

$$p.f = max(\omega) \tag{7}$$

$$s \cdot t : Crp_1 : \sum\nolimits_{a=1}^{m} P_a \leq P_{thr} \tag{8}$$

$$Crp_2 : \omega \in \{0,1\} \tag{9}$$

In this study, we set out to increase human life expectancy while keeping in mind the use of pesticides in the pertinent crop. The predetermined threshold value that limits the application of pesticides in the crop is represented by constraint Crp_1 in this case. Crp_2 indicates if it is 0 or 1.

4. The Proposed Framework

The AI-based system of classification presented in this section aims to increase human life expectancy by anticipating the pesticide threshold of a given crop. Different AI algorithms, including LSTM, MLP, SVM, as well as LR, were taken into consideration for the calculation of that threshold. Five levels make up the suggested framework: the edge intelligence layer, the data layer, the application layer, the blockchain layer as well as the communication layer.

Data layer

We have a variety of crops Crp_n in this stratum, including maize, rice, papaya, bananas, chickpeas, etc., in the farm f_i. Based on the maximum limit for the relevant crop, the farmer utilises the insecticides (threshold value) f_r. Typically, pesticides are used to eradicate plants and insects and increase the crop's productivity and yield. However, improper application of pesticides compromises human health and damages the soil, water, and turf [27]. As a result, we emphasised in the data layer that the farmer ffr_1 sprinkles pesticides p_m using an aerial vehicle or a farmer manually sprays the pesticide into the crop using a pesticide fountain.

$$\forall Crp_n \rightarrow p_1, p_2, \ldots p_{m,n} \in 1,2,3,\ldots \tag{10}$$

One comma-separated value (CSV) file contains the collected data from various farms f_i. The CSV file is then sent on to the edge intelligence layer for additional processing that precisely forecasts the pesticide threshold in the crop to maximise human life expectancy.

Edge Intelligence Layer

Data collection, preliminary processing, and probability prediction need a lot of computing resources in conventional AI methods. This is due to the fact that the input data (dataset) must travel to the global machines/servers for the reason of prognosis and classification, which lowers the computational efficiency of AI models. As a result, the input data doesn't depend on the global computers; rather, it is trained only on edge servers that have local AI training capabilities. This enhances the edge intelligence layer's effectiveness in terms of speedier response times and lowers the computational burden associated with data training.

Crop gathering information from the data layer in CSV file format is the first stage in creating this layer. To estimate the lifespan of humans, we have gathered the Kaggle [28] dataset. To train the AI systems, the gathered data is insufficient. As a result, we gathered information from additional sources that included threshold values and insecticides pertinent to the crop. The final dataset, D, shows how soil fertility is affected by temperature, humidity, and rainfall when different crop kinds are grown along with various pesticides and at their maximum permissible levels. The dataset consists of 12 columns $(CL_1, CL_2,...CL_k)$ and 900 rows $(R_1, R_2,...R_q)$.

$$R_1, R_2,...R_q \in D, q = 1,2,...900 \qquad (11)$$

$$CL_1, CL_2,...CL_k \in D, k = 1,2,...12 \qquad (12)$$

The attributes in dataset D have both categories as well as textual values. For instance, the label characteristics include the terms "Pesticide" as well as "Crop." The AI classifier only reads numerical input; it cannot process non-numerical data. In order to convert the categorical information into quantitative information and input it into the artificial intelligence (AI) system for future forecasts, we use the feature encoding E_n method.

$$\forall CL_g, CL_9 \xrightarrow{E_n} D \qquad (13)$$

The dataset undergoes pre-processing to remove missing data, duplicate entries, normalisation as well as noisy data, before being sent to the AI models. We identified the outliers that damage the dataset during data pre-processing and may result in incorrect predictions.

$$R_1, R_2, R_3 \in D \text{ and } R_4 \notin D \qquad (14)$$

For dataset D, we have R_1, R_2 and R_3. R_4 is, however, excluded from the dataset because it is not included in the dataset that takes into account the outlier. We use the Z-score method for normalisation and the box plot method to identify outliers in the suggested model. The Z-score normalisation method is illustrated in the diagram below.

$$\lambda = \frac{x_v - M}{std_v} \qquad (15)$$

where *std* is the number of standard deviations from the mean M. The results are then validated by splitting the pre-processed data into two partitions with a 70:30 split between training and testing. The training dataset Tr_D is the collection of information needed to train the classifier. The AI model Ml then uses the Tr_D to accurately forecast the pesticide threshold and examine how it would affect people's bodies. We employ a test dataset Ts_D that provides validation findings based on the training data after the model has been trained.

$$Tr_D \xrightarrow{fit} ML_\eta \xrightarrow{predict} \omega \begin{cases} 0 \\ 1 \end{cases} Ts_D \xrightarrow{fit} ML_\eta \xrightarrow{predict} \omega \begin{cases} 0 \\ 1 \end{cases} \qquad (16)$$

We used various AI-classifiers, including LSTM, MLP, SVM and LR. Based on the classification, we were able to get the best result from the LSTM classifier. Typically, the maximum marginal hyperplane is identified by the LSTM classifier, which splits the dataset into two classes. The hyperplane that best divides the two classes is generated in the first stage. In the subsequent stage, it chooses the hyperplane that appropriately separates the classes. LSTM is tuned using the hyper-parameters kernel as "rbf" and "linear," gamma as "auto," which utilises 1/n, where n is the number of features, and probability as "True," for the best results. Finally, based on the pesticide used on the crop, the AI model divides the data into two forms: high (high (0)) and low (low (1)) human life expectancy.

Depending on the classification of pesticide data, we project human life duration. Data security is necessary for human life expectancy data so that unauthorised users cannot change the data. To prevent an enemy from changing the threshold value and evading the food regulatory authority, the pesticide threshold data also needs to be validated and secured. As a result, we implement a blockchain network based on IPFS that offers authentication and secrecy against malicious data assaults.

$$\omega = \begin{cases} 0 & \text{humanl i f e } \uparrow \\ 1 & \text{humanl i f e } \downarrow \end{cases} \tag{17}$$

Blockchain Layer

A blockchain is a shared, unchangeable digital ledger made up of recordings of blocks. Multiple machines' transactions are recorded using blocks. There are four different kinds of blockchains: private, public, consortium, and hybrid. By safely storing them on the blockchain network, this layer offers security to the forecasted data, such as the human life expectancy that comes from the intelligence layer. We considered a distributed, permissionless public blockchain for that. These features imply that the public blockchain is not governed by a single entity and permits anyone to join the network to increase the transparency of the blockchain network. The attribute, "pesticide usage," which has an impact on human life expectancy, is the basis for the anticipated data in this case. The farmers, food providers, and retailers suffered a considerable loss of revenue if the projected data became public. As a result, the competitor can attempt to manipulate the anticipated statistics in order to increase sales of pesticide-based food products and maximise their profits. As a result, the agricultural ecosystem needs safe storage that can stop fraudulent operations. In order to put the anticipated data from the AI models into an immutable ledger that guards against data tampering and data injection attempts by malevolent users, we used an IPFS-based public blockchain. Every authorised user who has access to agricultural data thanks to public blockchain technology is fully informed of any changes made to the data that has been saved. Additionally, we store the data in IPFS using a smart contract called S_c that manages the decentralised network's high data storage costs. Faster content addressing, linking, and discovery are made possible by IPFS storage. Instead of identifying content by its location, it employs content addressing to determine what is inside of it. It employs directed acyclic graphs for content linkage, with each node having a distinctive identity that is a hash of the node's contents. Through the use of fixed-size hash tables, which enable effective data storage, it offers content discovery. This $hash_t$ table shares blocks of Ethereum with authorised users and adds new blocks to the network.

$$\omega \xrightarrow{\ S_c\ } IPFS \xrightarrow{\ generate\ } hash_t \tag{18}$$

$$hash_t \xrightarrow{\ add\ } \{Block_1, Block_2, \ldots Block_n\} \in Blockchain_D \tag{19}$$

The most well-known blockchain cryptographic hash function, SHA-256, which offers security through encryption, is used in this case. The following diagram represents the aforementioned object.

In the suggested framework, farmers, agricultural companies, food suppliers, and other authorised users can take part in securing agriculture-based information. It acts as a seamless link between government agencies and farmers to forecast the health consequences and life expectancy of pesticides. The blockchain copies in these nodes, which include farmers, suppliers, and government agencies, contain pertinent agricultural data from the intelligence layer. To improve the blockchain network's dependability and transparency, these copies are sent to every node. In order to promote food products with the least amount of pesticides, authorised users can obtain secure agricultural data from the blockchain nodes and send it to the application layer.

Application Layer

This layer consists of several government agencies and organisations that use agricultural data for additional investigation and study. For instance, if a company wishes to examine the impact of pesticides on human health, it can obtain the information from the repository. Data on human life expectancy can be used to determine the toxicity of insecticides. Infants as well as young kids are especially susceptible to the effects of insecticides. It has mysterious effects on the human body, including nausea, eye-stinging, blisters, skin rashes, dizziness, blindness, and other long-term effects.

Communication Layer

The relationship between all four layers is included in this layer. The farmers spray pesticides manually or with drones on the farms depicted in the data layer. Massive amounts of data must be gathered and processed in order to implement the intelligent edge layer. Additionally, using tablets, laptops, smartphones, or any other 6G-based device, the government agency or organisation retrieves the pertinent data at the application layer. Each one of them needs a reliable communication path with a high data throughput and low latency. Hence, using a 6G interface is necessary to obtain safe and pertinent data for analysis. The 6G network that the smart agriculture application utilises has a predicted spectrum range of 110 GHz up to 170 GHz and a maximum throughput of 1 Tbps [29–30].

5. Results and Discussion

This section examines the effectiveness of the suggested AI-based smart agricultural framework, which forecasts the optimal levels of pesticide use by taking into account several AI models. The following is a thorough analysis of performance evaluation.

Experimental Setup

To develop and assess the suggested framework, many software and application platforms are used. First, the AI intelligence is put to use in the anaconda distribution, which includes a Jupyter notebook and a number of libraries for visualisation, data pre-processing and training including Scipy, Sklearn, Python, Numpy, Pandas, and Matplotlib. The data that was accurately anticipated is then sent to the blockchain layer that is based on IPFS and is implemented in the Remix "integrated development environment (IDE)" . The Remix IDE provides an online platform for building custom smart contracts using a variety of solidity-based functions, including getfarmers() to retrieve a farmer's identity, Registerfarmer to add a farmer to the smart contract, Checkwitness() to confirm the registration of a farmer, Pesticideuse() to check for pesticide use on a crop, addcrop(), addpesticide(), getTransactioncount(), and many more. Truffle is used to create the smart contract, which is then released across the Ganache network. Only the verified data is sent to the IPFS-based blockchain ledger after the smart contract has been completed. It transforms the data into hashed data that can be easily fetched from the blockchain network, improving the proposed framework's capacity to scale. Additionally, the 5G toolbox is used to mimic the 6G network interface for manipulating the 6G network in the matrix laboratory, or MATLAB 2022a. This is possible by changing a number of 5G toolbox characteristics, including frequency range, channel bandwidth, subcarrier spacing, modulation types, etc.

To enhance the effectiveness of AI training, various AI models and their hyper-parameters are used in the suggested work. In this case, the decision tree that makes decisions by dividing nodes into sub-nodes is split using the "best" and "random" splitters. Moreover, the population is divided into two halves using the "gini" criterion.

It is determined by deducting one from the total of the squared probability for each class. For the LSTM classifier, we employed the gamma, probability, and kernel hyperparameters that Low values indicate "far" and high values indicate "close" in the gamma parameter, which measures the influence of a single training set. Gamma 'auto' in this context denotes the utilisation of 1n features. To control whether probability estimations are enabled, a probability hyperparameter is employed. The rbf kernel was chosen from among the LSTM's "linear," "poly," "rbf," "sigmoid," and "precomputed" kernel types because it offers shortcuts for avoiding difficult calculations. Estimators of 10 were employed by the random forest classifier to represent the total number of trees in the forest. Alphafloat is a constant that multiplies the regularisation term, while l1_ratiofloat is the elastic net mixing parameter with a 0.15 value. These two hyper-parameters were employed by Perceptron.

Dataset Description

The Kaggle agricultural dataset includes information on crop fertilizer, rainfall, and climate for the Indian region. The dataset is simply insufficient to train the AI model, though. In order to train the AI to discover comparable datasets, we looked into various online resources such as legitimate blogs, websites that compete in artificial intelligence, and academic papers. Finally, we combined all of the records into a single dataset with the desired properties, such as the maximum limits for pesticides, environmental variables, crop and soil conditions, and environmental factors. The data was then pre-processed before being added to the design that reliably forecasts the threshold of pesticides to maximise the life span of humans. Encoding was then used to manage the numerical numbers.

Experimental Analysis

The outcome states that the analysis of the recommended smart agricultural framework is presented in this section. We took into account the dataset with the effects of humidity, temperature, pH, rainfall, as well as pesticides with their highest cut-off value for the corresponding crop. The balance 30% testing set is utilised to evaluate the training set, in which we sent 70% of the training data into the AI algorithms. Once after the effective training phase, the AI algorithms predict the pesticide threshold for the specific crop.

The pesticide usage data shown in Fig. 1 depends on the previously defined maximum cut-off value for the corresponding crop. With the extension of the population comes an increase in the need for agricultural goods. Farmers apply

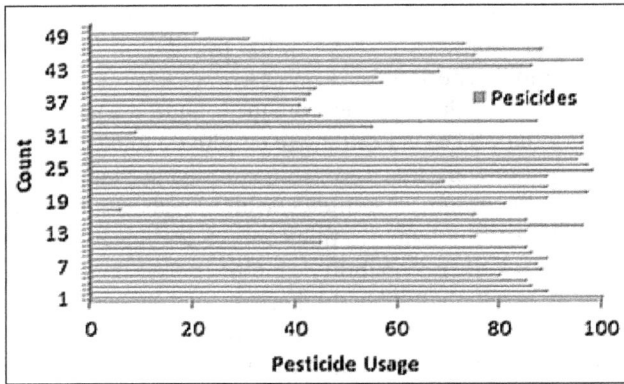

Figure 1. Use of pesticide in crop.

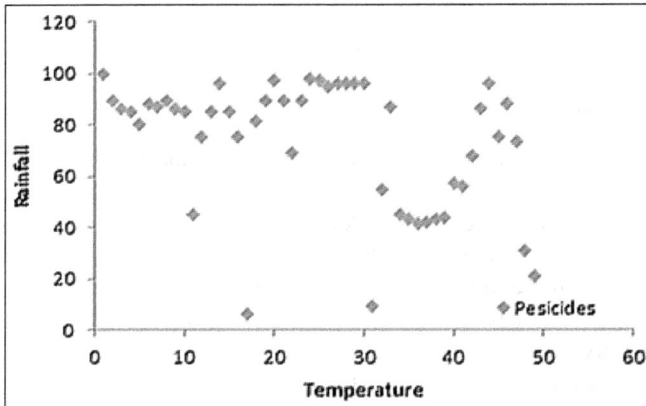

Figure 2. Temperature and rainfall effect on crop.

pesticides that improve the land's productivity and futility in order to meet the demand. The AI algorithms can profitably export additional agricultural products to the market as a result. The total number of insecticides utilised by farmers for their crops is depicted in Fig. 1. The y-axis shows the thresholds of the pesticides used by farmers, while the x-axis shows the number of pesticides used. For instance, farmers who wanted to increase crop profits used pesticides no more than 125 times at a threshold of 10, and no more than 26 times at a threshold of 20. The human body is harmed if the pesticide (P) value exceeds the pesticide threshold (Pthr), and vice versa. In addition, we examined how humidity, pH, rainfall and temperature affect crop output. The crop's response to temperature and rainfall is depicted in Fig. 2, which lowers crop quality and renders the land ineffective. Farmers occasionally are unable to harvest their crops in a timely manner due to temperature and rainfall. Therefore, to increase the pace of agricultural production growth, farmers apply a variety of pesticides.

The "Long and Short Term Memory (LSTM), Multi-Layer Perceptron (MLP) SVM, and Logistic Regression (LR)" is used by the cognitive layer to train the pesticides-based agriculture dataset. We took into account a number of performance metrics, including precision, accuracy, receiver operating characteristic curve (ROC), loss, F1-score, and recall to assess the intelligence layer's performance. To do so, the values of the abovementioned metrics— true positive ($true_{pr}$), true negative ($true_{ng}$), false positive ($false_{pr}$), and false negative ($false_{ng}$)—are determined by computing the confusion matrix for each algorithm. Here, accuracy refers to the proportion of observations out of all the observations that are accurately predicted as positive. Accuracy alone, however, is insufficient to identify the finest AI programme. As a result, we used precision and recall to support the accuracy metric's findings. The ratio of the genuine positive to all other positive outcomes ($true_{pr}$ + $false_{pr}$) is specified by the accuracy value. The ratio between all instances of true positive and false negative is defined by the recall value. This indicates that from both positive and negative results, what proportion is appropriately labelled as positive. The following is how these measures are expressed mathematically:

$$M_{acc} = \frac{true_{Pr} + true_{ng}}{true_{Pr} + true_{ng} + false_{Pr} + false_{ng}} \tag{20}$$

$$M_{pre} = \frac{true_{Pr}}{true_{Pr} + false_{Pr}} \tag{21}$$

$$M_{recall} = \frac{true_{Pr}}{true_{Pr} + false_{ng}} \tag{22}$$

Additionally, we may modify the f1 rating as an additional performance metric that takes into account both recall as well as precision values for calculating the f1 score if both precision and recall values are suboptimal and exhibit less importance.

We discovered that the LSTM beats other AI algorithms in terms of recall, accuracy, f1 score, and precision after examining all the values of performance criteria. Figure 3 compares the accuracy of various AI algorithms, including LSTM, MLP, SVM and LR. Based on the empirical findings, we discovered that decision trees have lower accuracy, precision, and recall scores because they are heavily biased towards the training set and have a significant likelihood of overfitting. A random forest is also computationally demanding because it collects vast amounts of tree samples and then calculates the probability for each sample. As a result, it becomes sluggish and useless for making predictions in real-time. Additionally, the perceptron's complicated

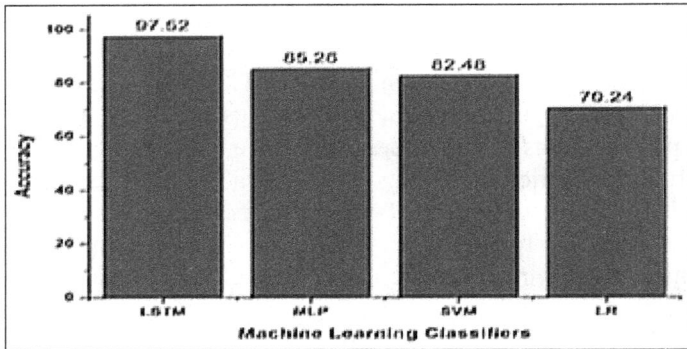

Figure 3. Accuracy of AI algorithms.

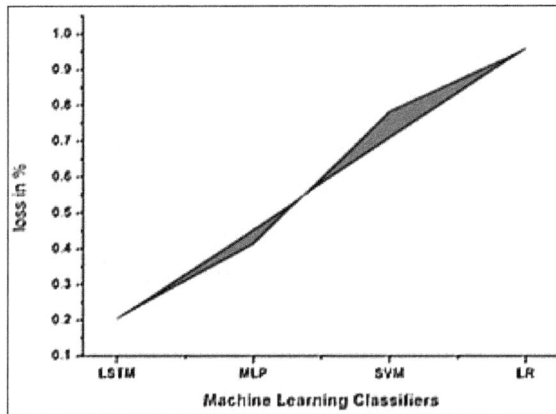

Figure 4. Loss score of AI algorithms.

transfer function makes it inappropriate for use with nonlinear data values. In contrast, LSTM performs better than other classifiers because it uses effective kernel functions to support both non-linear as well as linear solutions. After determining the optimal plane with the greatest margin between discrete data values, such as 0 (high human life expectancy) or 1, it swiftly converges. We discovered from the thorough study of the findings that the LSTM method outperforms the MLP, SVM and LR in terms of accuracy as shown in Fig. 4. For the provided dataset, it offers 97.63% training accuracy and 96% testing accuracy. This finding demonstrates that the AI model being used 96% accurately forecasts human life expectancy based on the pesticides of relevant crops that are being provided.

Table 1 compares AI models using various performance criteria, including recall, accuracy, precision, and f1-score. A log-loss score graph of the various AI classifiers, which measures how well the classifiers work, is shown in

Table 1. Comparison of AI algorithms with different metrics.

	Precision	F1score	Accuracy	Recall
LSTM	94.26	96.45	97.52	98.02
MLP	83.12	82.77	87.26	84.56
SVM	81.44	76.35	82.48	70.22
LR	71.26	70.44	70.24	69.84

Fig. 2. It shows how well the expected values for human life expectancy match the actual values. We deduce from the graph that perceptron and random forest have larger log-loss scores due to greater divergence between the predicted and observed values. The log-loss score for the LSTM and decision tree, i.e., 0.234 and 0.392, respectively, is smaller for these two methods since their accuracy is higher than that of random forest and perceptron. The thorough outcome demonstrates that the LSTM classifier improved the performance of the suggested framework by precisely projecting the life expectancy of humans based on pesticide usage. Additionally, an effective protocol built on a blockchain called IPFS is employed for safe data storage. Each user operates a node in the blockchain network, and nodes communicate with one another to exchange data. However, managing data requests on the conventional blockchain is laborious. Therefore, the suggested framework makes use of IPFS's incredible advantages within the blockchain network. In comparison to traditional blockchain technology, it has a number of advantages, including the capacity to hash data and store it in an immutable ledger, increase scalability, and lower the cost of data storage. In contrast to the traditional blockchain network, the stored data (predicted data) can be fetched easily and quickly thanks to the IPFS's use of hashing. This shows that the IPFS-based blockchain has a faster reaction time than the conventional blockchain, which stores raw data. A rapid response time allows the IPFS-based blockchain to accommodate concurrent data requests effectively, which contributes to the proposed framework's excellent scalability.

6. Conclusion

By using important enabler technologies (such as artificial intelligence (AI), blockchain, etc.), the smart agriculture industry contributes significantly to the growth and improvement of the country's economic success. However, challenges related to More number population and change in climate are raised in the field of the agriculture industry. One of the biggest problems is inorganic farming, which shortens people's lives. In this study, we developed a smart agriculture system based on blockchain and AI that accurately forecasts the life of human expectancy depending on the utilisation of pesticides in food products. In order to achieve this, we first acquired agricultural pesticide statistics from a variety of online sources, including websites that compete in artificial intelligence and academic articles. The dataset is then pre-processed to enhance the effectiveness of data training utilising data pre-processing techniques. Additionally, AI models like LSTM, SVM, MLP, and LR are utilised for accurate human life prophecy, i.e., high or low, depending on the intake of pesticides. The public blockchain network powered by IPFS houses the accurately anticipated agricultural data, enhancing the security and privacy of the suggested architecture.

Additionally, adding a 6G network interface enhances interaction between the framework's many parts. Finally, the performance of the proposed method parameters—scalability, packet loss ratio and accuracy are assessed. The outcome depicts that the LSTM performs better compared to the other AI algorithms by attaining a higher prediction accuracy of nearly 97.63% with 62.13 ms; latency and 23.54% of minimal packet loss ratio.

By thoroughly examining contemporary security risks including hacking and routing assaults in blockchain as well as the proposed smart AI-based agriculture, the security implications of the proposed model can be extended as further work.

References

[1] Abdullah, A., Enazi, S.A. and Damaj, I. 2016. AgriSys: A smart and ubiquitous controlled-environment agriculture system. pp. 1–6. *In*: 2016 3rd MEC International Conference on Big Data and Smart City.

[2] Nations, U. 2022. Peace, dignity and equality on a healthy planet. Online https://www.un.org/en/global-issues/population [Accessed 20 May 2022].

[3] Smart Farming powered by analytics. 2022. https://www.wipro.com/analytics/smart-farming-powered-by-analytics/ [Accessed 27 May 2022].

[4] Global, S. 2022. Modern agriculture has many complex challenges. https://www.syngenta.com/en/innovation-agriculture/challenges-modern-agriculture [Accessed 21 May 2022].

[5] What is the difference between fertilizer derived from organic and synthetic sources? 2022. https://www.milorganite.com/lawn-care/organic-lawncare/organic-vs-synthetic [Accessed 27 May 2022].

[6] Ahmed, U., Lin, J.C.W., Srivastava, G. and Djenouri, Y. 2021. A nutrient recommendation system for soil fertilization based on evolutionary computation. Comput. Electron. Agric 189: 106407.

[7] Hassaan, M.A. and Nemr, A.E. 2020. Pesticides pollution: Classifications, human health impact, extraction and treatment techniques. Egypt J. Aquat. Res. 46(3): 207–20.

[8] Sakthi, U. and DafniRose, J. 2022. Blockchain-enabled smart agricultural knowledge discovery system using edge computing. Procedia Comput. Sci. 202: 73–82, International Conference on Identification, Information and Knowledge in the Internet of Things, 2021.

[9] Feng, Y. 2022. Application of edge computing and blockchain in smart agriculture system. Math Probl. Eng. 2022.

[10] Patel, N., Shukla, A., Tanwar, S. and Singh, D. 2021. KRanTi: Blockchain-based farmer's credit scheme for agriculture-food supply chain. Trans Emerg. Telecommun. Technol. doi:https://doi.org/10.1002/ett.4286.

[11] Jiao, L., Dong, S., Zhang, S., Xie, C. and Wang, H. 2020. AF-RCNN: An anchor-free convolutional neural network for multi-categories agricultural pest detection. Comput. Electron Agric 174: 105522.

[12] Sathiyaprasad, B., Seetharaman, K. and Kumar, B.S. 2020. Content based video retrieval using Improved gray level Co-occurrence matrix with region-based pre convoluted neural network–RPCNN. pp. 558–563. In 2020 3rd International Conference on Intelligent Sustainable Systems (ICISS), IEEE.

[13] Kumar, B.S., Seetharaman, K. and Sathiyaprasad, B. 2022. Content-based video retrieval based on security using enhanced video retrieval system with region-based neural network (EVRS-RNN) and K-Means classification. pp. 401–414. In International Conference on Artificial Intelligence and Sustainable Engineering: Select Proceedings of AISE 2020, Volume 2, Singapore: Springer Singapore.

[14] Chen, C.J., Huang, Y.Y., Li, Y.S., Chang, C.Y. and Huang, Y.M. 2020. An AIoT based smart agricultural system for pests detection. IEEE Access 8: 180750–61.

[15] Menaka, D., Gauni, S., Indiran, G, Venkatesan, R. and Muthiah, M.A. 2022. A heuristic neural network approach for underwater parametric prediction at Bay of Bengal. IETE Journal of Research (2022): 1–10.

[16] Menaka, D. and Gauni, S. 2022. An energy efficient dead reckoning localization for mobile Underwater Acoustic Sensor Networks. Sustainable Computing: Informatics and Systems 36: 100808.

[17] Menaka, D., Gauni, S., Indiran, G., Venkatesan, R. and Muthiah, M.A. 2022. Development of heuristic neural network algorithm for the prognosis of underwater ocean parameters. Marine Geophysical Research 43(4): 40.

[18] Priyadharshini, A., Chakraborty, S., Kumar, A. and Pooniwala, O.R. 2021. Intelligent crop recommendation system using machine learning. pp. 843–8. In: 2021 5th International Conference on Computing Methodologies and Communication.

[19] Nalwanga, R., Nsenga, J., Rushingabigwi, G. and Gatare, I. 2021. Design of an embedded machine learning based system for an environmental-friendly crop prediction using a sustainable soil fertility management. pp. 251–6. In: 2021 IEEE 19th Student Conference on Research and Development (SCOReD).

[20] Deivakani, M., Singh, C., Bhadane, J.R., Ramachandran, G. and Kumar, N.S. 2021. ANN algorithm based smart agriculture cultivation for helping the farmers. pp. 1–6. In: 2021 2nd International Conference on Smart Electronics and Communication.

[21] Jin, X., Che, J. and Chen, Y. 2021. Weed identification using deep learning and image processing in vegetable plantation. IEEE Access 9: 10940–50.

[22] Kanuru, L., Tyagi, A.K., Aswathy, S.U., Fernandez, T.F., Sreenath, N. and Mishra, S. 2021. Prediction of pesticides and fertilizers using machine learning and internet of things. pp. 1–6. In: 2021 International Conference on Computer Communication and Informatics.

[23] Kasinathan, T., Singaraju, D. and Uyyala, S.R. 2021. Insect classification and detection in field crops using modern machine learning techniques. Inf. Process Agric. 8(3): 446–57.

[24] Sreedevi, B., Mohanraj, G., Revathy, J. and Roobini, R. 2022. Agri brilliance-a farm log rental service platform with crop and disease management using machine learning techniques. pp. 1–7. In: 2022 International Conference on Advances in Computing, Communication and Applied Informatics.

[25] Elakkiya, E. and Karthik, P. 2022. Evaluation on correctness agriculture—soil quality and soil borne disease in india using machine learning. pp. 1–6. In: 2022 International Conference on Advances in Computing, Communication and Applied Informatics.

[26] Durai, S.K.S. and Shamili, M.D. 2022. Smart farming using Machine Learning and Deep Learning techniques. Decis Anal. J. 3: 100041.

[27] Setiawan, A., Yudistira, N. and Wihandika, R.C. 2022. Large scale pest classification using efficient Convolutional Neural Network with augmentation and regularizers. Comput. Electron. Agric 200: 107204.

[28] Gopikrishnan, S., Srivastava, G. and Priakanth, P. 2022. Improving sugarcane production in saline soils with machine learning and the internet of things. Sustain Comput: Inf. Syst. 35: 100743.

[29] Ingle A. Crop recommendation dataset. 2022. https://www.kaggle.com/datasets/atharvaingle/crop-recommendation-dataset [Accessed 12 May 2022].

[30] Chowdhury, M.Z., Shahjalal, M., Ahmed, S. and Jang, Y.M. 2020. 6G wireless communication systems: applications, requirements, technologies, challenges, and research directions. IEEE Open J. Commun. Soc. 1: 957–75.

CHAPTER 6
Reversible Data Hiding
Methods and Applications in Secure Medical Image Transmission

Shaiju Panchikkil and *V. M. Manikandan**

1. Introduction

Images of the human body can help doctors in diagnosing the actual problem and hence offering an efficient treatment plan to the patient. Very common imaging techniques include X-rays, CT (Computed Tomography) scans, MRI (Magnetic Resonance Imaging), Nuclear Medicine Imaging (Positron Emission Tomography), and ultrasound imaging. The Covid-19 pandemic has affected humankind in one way or the other. In fact, getting treatment is inevitable, but exposing our-self to the public while visiting a doctor or a hospital can make us prone to catching Covid-19. Even health practitioners had difficulties in treating patients during this pandemic. Different types of technologies have been playing a great role upfront to combat all the constraints between a health practitioner and the patient. The future of healthcare will be molded by the advancement of digital healthcare technology. As the digital era continues, digital communication also expands, and thus medical image communication. Transmission of digitised medical images through a communication channel is risky, because:

- The communication channel itself is not secure.
- The medical image transmitted contains confidential private information of the patient.
- Any alterations to the image can mislead a proper diagnosis and can result in erroneous treatment.

SRM University-AP, Andhra Pradesh, India.
* Corresponding author: manikandan.v@srmap.edu.in

Thus, securing the digitised medical image is a non-trivial matter. There are different formats for representing a medical image viz., Analyse, Neuroimaging Informatics Technology Initiative (Nifti), Minc, and Digital Imaging and Communications in Medicine (Dicom). Analyse image uses .img and .hdr as extensions, Nifti image uses .nii as an extension, Minc image uses .mnc as an extension and Dicom images use .dcm as the file extension. DICOM has been widely accepted as a standard for storing, retrieving, and transmitting medical information. A medical image when represented in a computer or any other device in a digital form would consist of the basic element called a pixel. Each of these pixels indicates a piece of medical information and hence the pixels in the totality of an image specify the size of the image. A 16-bit Dicom image implies that each pixel is represented by 16 bits of information.

A Dicom file contains a Preamble, which indicates its compatibility with other file formats, a Prefix that contains the string 'D', 'I', 'C', 'M', and a Header that contains the metadata and image Dataset. Here the metadata would signify the information about the image. In a certain context, the preamble and prefix are also referred to as part of the header. In medical images, the header part of the information plays a major role. The image dimensions, number of bits per pixel (bpp), resolution, machine settings while capturing the image, demographic data, study parameters, etc. are the information stored in the header. Hence metadata can be exploited to gain image-related information for clinical as well as research purposes.

There are two dimensions to the research activities associated with medical images. The first dimension of the research activity concerns securing the image contents via the Reversible Watermarking technique and another activity is where the additional metadata information is added within the original medical image via the Reversible Data Hiding technique. Both of these techniques will be discussed in detail in the section on related works.

2. Data Hiding—An Overview

Cryptography was an initial method of secure communication wherein the original information as such is converted into an unreadable form called cipher data before transmission. This scrambled presentation of the information can attract a third party's attention and may lead to attacks. Data hiding emerged as a technique of secure communication where the original information is not modified but the information here is embedded inside another cover medium for communication. The advantage of the data hiding technique over cryptography is that the actual information is hidden within a cover medium and hence no one can identify the presence of confidential data on viewing the cover medium. This can eventually reduce the probability of attack to

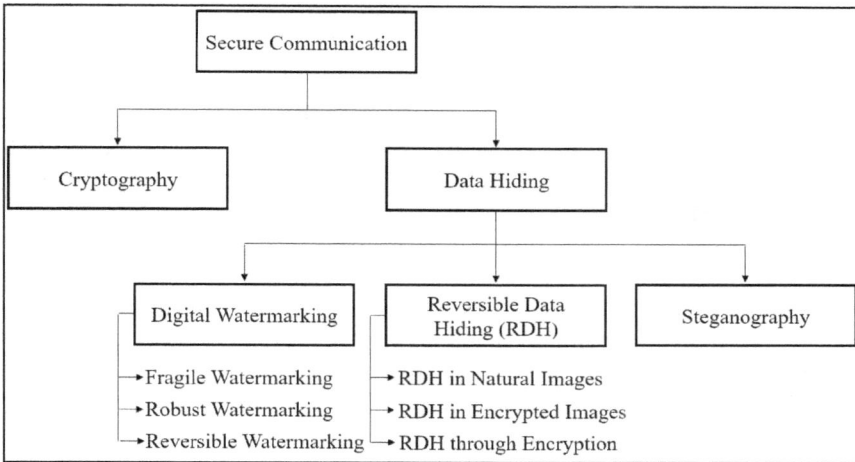

Figure 1. Secure communication.

gain knowledge on the hidden information. The basic hierarchy of secure communication is shown in Fig. 1.

In the data hiding technique, the cover medium acts as a carrier for holding the original information. The different types of cover mediums are image, text, audio, and video.

Digital watermarking is a technique in which a unique sequence of data is placed in the cover medium. This unique information could be in the form of a logo, an image, a text, or a digital signature. Watermarking is a very useful technique. Researchers are immensely keen to explore watermarking techniques, to mould and expand its capability to improve the secure communication of confidential data. The watermarking capability can be exploited in two different ways: Fragile watermarking and Robust watermarking. If the extracted watermark from the cover medium is used to detect any changes to the original work, then the method is named fragile watermarking. It's a very sensitive type of watermarking, used for tamper detection or protecting data integrity. Whereas, the case of using the embedded watermark to combat geometric interpolations like rotation, scaling, shearing, etc., and general operations like distortion, compression, etc. are marked as robust watermarking. This type of watermarking is mainly used for ownership protection or copyright protection. Reversible watermarking has emerged as a solution to authenticating the contents and security of medical images. Inserting a watermark in a medical image will distort the image and thereby influence the physician's diagnosis which is intolerable. Hence, using a watermark in medical images should comply with removing the hidden watermark and thus restoring the original image. Reversible watermarking has the capability of reproducing the patient's private data without a loss after removing the watermark.

Reversible Data Hiding (RDH) is another form of data hiding mechanism where the metadata or the Electronic Patient Record (EPR), which is highly confidential will be embedded within a cover medium and communicated. It is noteworthy to point out that, unlike conventional data hiding schemes where the cover medium gets modified during the process of hiding the secure information and the receiver is unable to recover the cover medium, reversible data hiding promises complete reversibility by guaranteeing the recovery of the original cover medium and the entire hidden confidential information. Hence, RDH is one of the most promising techniques which can be utilised in highly sensitive applications like the transmission of EPRs. There are three different classifications of RDH with respect to an image as a cover medium, i.e., using natural images, using encrypted images, and RDH through encryption.

An RDH communication model consists of three elements: the cover medium owner, the information concealer, and the receiver. The cover medium owner is the possessor of the cover medium. The information concealer is responsible for embedding the confidential data inside the cover medium and the receiver is the entity with whom the information is being communicated. RDH in natural images uses the natural image as the cover medium and the medium owner transfers the original image directly to the information concealer. Thus, the information concealer embeds the additional information over the original image. It should be noted that the information concealer gains the details of the cover medium before the embedding process. Whereas, in the case of an RDH in encrypted images, the cover medium owner secures his cover image by encrypting the cover image before transferring it to the information concealer. Hence, the information concealer will not have any idea about the cover image during the information embedding process. The third kind of RDH, termed RDH through encryption, is very similar to RDH in encrypted images, where the sending party is a single entity, i.e., the cover medium owner and the information concealer are treated as a single person like a doctor who doesn't want to disclose any details of his patient to anyone else other than the recipient doctor.

Steganography is also an information hiding technique, where the secret information is concealed within another cover medium that is not secret. The main objective of steganography is to extract the secret data without any loss at the receiver end, i.e., the resulting cover medium after the extraction of the secret data is of the least significance.

Recent Related Works

This section gives insight into the recent related works on reversible watermarking and reversible data hiding in medical images [18] that have been well-acknowledged by the research community.

Related Works on Reversible Watermarking

Every medical image before applied to any operation is subjected to partitioning into the region of interest (ROI) and region of non-interest (RONI). ROI is the portion of the medical image that is very important, highly sensitive, and directly impacts the physician's diagnosis. Hence, these portions should not be altered by any cause and need reversibility. RONI holds the patient's details and other information which needs to be authenticated. So, the robustness of the system can be achieved at RONI. Very often the programmer or the designer of the system chooses what to achieve and accordingly sets the parameters. Reversible watermarking schemes can be implemented via various approaches viz., compression [6, 8, 36, 19], difference expansion [34, 39, 28, 13, 15], histogram shifting [12, 22, 25, 14, 21], and transformation based [33, 17, 41]. A few of these research works are discussed below.

A genetic algorithm (GA) and integer wavelet transform-based reversible watermarking scheme is proposed in [6]. Here a threshold matrix is generated via the GA algorithm that helps in the efficient embedding of the watermark. Before watermark embedding, to overcome the overflow and underflow issues, the histogram of the medical image gets modified, i.e., for a grayscale image, the normal gray range is [0–255]. It becomes [1–254] after modifying the histogram. A scan sequence is also maintained to reinstate the normal histogram during the extraction process. The image hence obtained is converted in wavelet sub-bands by employing integer wavelet transform. HL, HH, and LH are selected from the generated sub-bands for watermarking purposes. The proposed scheme could improve the watermark payload and quality of the recovered image.

A Recursive Dither Modulation (RDM) based reversible watermarking scheme is discussed in [15]. In this scheme as well, wavelet transformation of the image is performed to get the low-frequency wavelet coefficient which is used for embedding the watermark. The strength of the watermark is controlled by quantisation steps defined via differential evolution to better the robustness. A difference expansion-based information embedding scheme that utilises predictors for predicting the pixel values is explored by [20]. In the difference expansion method, similar pixels form a group and additional data is added to their difference. The proposed technique also uses the prediction error method where neighbour pixels support the prediction of cover image pixels. Hence, prediction error acts as a means to embed a bit from the watermark through difference expansion. If (i, j) indicates the location of a pixel $Z(i, j)$ of the cover image and the predicted value is given by $Z'(i, j)$ then the prediction error is found using the formula:

$$Xerror = Z(i, j) - Z'(i, j)$$

The distortion effect caused via embedding of the watermark is reduced by setting a threshold. If T indicates the threshold, then the watermark bit Z1 is embedded and transformed into a new pixel value by:

$$Xnew = Xerror + Z(i, j) + Z1, \ if\ Xerror < T$$

The watermarked image along with other information like the prediction image and threshold is transmitted to the receiver where the recovering process is carried out. In this scheme, the prediction methods used were median prediction to predict the border pixels and least square prediction to predict other pixels. At the receiver side, the image with the watermark and the prediction image are utilised to calculate the prediction error. The simulation results reveal that the increase in embedding capacity has introduced more distortions into the cover image, thereby reducing the peak signal-to-noise ratio (PSNR). When the recovered image is exactly of the same quality as that of the cover image, PSNR will be ∞.

A reversible watermarking system using wavelet decomposition and particle swarm optimization (PSO) technique is proposed by [7]. Here the medical image is initially transformed using discrete wavelet transformation and a watermark is inserted by altering the wavelet coefficients. The watermark image hence generated is treated with a chaotic tent map and encrypted. Tent maps are usually efficient for image encryption and video transmission. Each watermark bit is added to the HL sub-band component and whose fitness was supported by PSO. Optimal wavelet coefficients are found to facilitate the hiding of the watermark bits. The proposed scheme successfully embedded a watermark with low distortion. Research from [29] proposed a fully reversible watermarking scheme that combines Reversible, Zero, and RONI watermarking approaches. Here the zero watermarking technique is a lossless method as it doesn't modify any data, where the watermark is not embedded but kept for later comparison. This is applied to the ROI area combined with dual-tree complex wavelet transform (DT-CWT). DT-CWT will generate a binary matrix related to the LL subband of DT-CWT coefficients. An embedding greater than 0.04 watermark bits per pixel is achieved by the proposed method that is based on reversible contrast mapping (RCM). However, the scheme requires the location map to extract the embedded information and recover the cover image. Another reversible watermarking scheme for DICOM MR images is introduced in [27]. It is a difference expansion-based approach and the technique is capable of detecting all kinds of manipulations in the medical image. Here the image gets segmented into ROI and RONI. Each of the segments is processed as blocks of size 3 × 3 pixels. Watermark bits are embedded inside the 3 × 3 pixel smooth blocks of ROI. Smooth blocks are those blocks with pixels having very low intensity differences. Selecting smooth blocks can achieve the goal of imperceptibility on modification. In this scheme, the level of modification that applies to the medical image is evaluated via trials using Visual Grading Analysis (VGA) to bring down the modifications to be visually perceptible. Another peculiarity of the scheme is the extraction of the watermark without using a location map, which enhances the embedding capacity. Two types of watermarks were used in the scheme to ensure authentication and to offer integrity. The authentication watermark is generated by utilizing the patient information and constant image data from the DICOM header. The digital signature defined as an integrity watermark is generated by applying the message digest (MD5) algorithm to the medical image. These two watermarks are concatenated and transformed into binary form. Watermark is compressed using the run-length encoding technique. The proposed scheme could successfully embed a watermark of 4 bits within each 3 × 3 pixel block. The authenticity of the image data and integrity can be ensured by extracting these two watermarks.

The literature survey of the related works on reversible watermarking in medical images reveals that the areas of focus in research works have been:

- Capability to verify the authenticity and integrity of DICOM image.
- Embedding watermarks into RONI.
- Embedding watermarks into ROI.
- Complete removal of the watermark after detection and verification to restore the original pixel values.
- Increasing the embedding capacity.
- Improving the image quality.
- Lowering complexity.

Related Works on Reversible Data Hiding

The reversible data hiding technique demands recovery of the cover image without any quality loss along with the extraction of all embedded metadata. Due to this property, reversible data hiding has been explored to make it useful in highly sensitive applications like the transferring of EPRs along with the clinical image where no permanent modifications are tolerable. Reversible watermarking has been classified into different approaches based on the way to create space for embedding the watermark, similarly, RDH-based schemes can also take different approaches like difference expansion [10, 34], histogram shifting [22, 32], and prediction error based [26, 23].

An RDH for medical images with the intention to enhance the contrast is proposed in [35]. Initially, a background segmentation is done to identify the principal grey levels. The segmentation is done by selecting an optimal threshold to separate the ROI and RONI. Since it's not worthy of enhancing the RONI contents, the histogram of foreground segments is expanded by controlling the overflow and underflow problems. The technique used here is histogram shifting. Modifications on pixels are tracked by maintaining a location map. The histogram bins with fewer pixels are selected adaptively from the ROI. This is to reduce the visual distortions introduced during the pre-processing stage. Now those histogram bins which belong to the foreground segment are used for embedding the metadata. Hence, the contrast of ROI is improved while retaining the background contract of the medical image. A reverse process is performed wherein the embedded metadata gets extracted and the cover image also gets recovered. The results show that the visual quality has improved. However, the histogram pairs counted and selected for expansion should not exceed 60 to maintain a tolerable visual distortion.

The RDH schemes can also be applied to encrypted medical images. One such scheme is discussed in [16]. Here the given 16-bit grayscale medical image is classified into ROI, RONI, and border regions. ROI is chosen by the content owner by defining a polygon. Vertices of the polygon are preserved. A hash value of the ROI is calculated using the MD5 hashing algorithm for authentication purposes. Now the image is rearranged placing ROI in front, followed by RONI and border region pixels. The encryption key is used to encrypt this image matrix. Metadata and hash values are hidden by utilising the least significant bits (LSBs) of this encrypted image. In this case, as the metadata and hash value are embedded in an encrypted image, the privacy of the contents of the original medical image is preserved.

Another RDH with an objective of contrast enhancement along with the capability of tamper localisation for medical images and no distortion in the ROI is proposed in [12]. It follows a similar process of implementation as seen earlier. Initially, the segmentation is performed to separate the background and ROI using an automatic optimal thresholding algorithm. In the pre-processing stage, histogram shifting is used to enhance the contrast of ROI. For this peak pairs from the histogram need to be adopted. If A is the peak pair and R_z is the minimum pixel from ROI, then all pixels in $[R_z, R_z + A - 1]$ are increased by A and those in $[256 - A, 255]$ are decreased by A, i.e., here the background pixels will lie within [0, $R_Z - 1$] and ROI pixels lie within [R_Z, 255]. The conditions for the adopted peak-pairs A ϵ [1, R_T] and should satisfy

$$R_T = \left\lfloor \frac{(R_A - R_Z)}{2} \right\rfloor$$

$$R_A = \left\lfloor \frac{(255 - R_Z + 1)}{2} \right\rfloor$$

A location map is maintained to note the pixel locations that contain a minimum number of pixels within the two intervals. After finding the histogram of the ROI without considering the first 16 pixels, pair of peak bins are identified as P_Z and P_R such that $P_Z < P_R$. Now the metadata is embedded by using the procedure: Pixels X lesser than P_Z are decremented, Pixels X greater than P_R are incremented, Pixels in $P_Z < X < P_R$ is retained as such, pixels equal to P_Z are decremented by metadata bit and pixels equal to P_R is incremented by metadata bit. Now a mask is generated from the image by setting the background pixels as 0. This mask is processed as independent 4 × 4 pixel blocks. A two-level lifting wavelet transform is made over the mask to retrieve a low-frequency component of size 1 × 1. Combining the frequency sub-bands to generate a low-frequency coefficients image and its fifth-bit plane is used as a feature-bit matrix. Arnold transforms of this fifth-bit plane matrix is created for improving the security further. The minimum pixel value of ROI R_Z and the number of feature bits Z_f are stored in the first 23 LSBs of background pixels. In a due manner, the next Z_f LSBs will hold the scrambled feature bits. The receiver tries to extract this scrambled feature-bit matrix to understand the ROI feature and for tamper localisation.

Earlier the metadata was embedded in smooth areas for attaining a high Peak Signal-to-Noise Ratio (PSNR) but the method proposed in [37] uses the texture region of medical images for hiding metadata. This helps in having a better detailing of the information in the image for accurate diagnosis. A message sparse representation has also been introduced to reduce the distortion introduced while enhancing the contrast of texture region and to improve the embedding efficacy. Histogram stretching is the technique used in the proposed scheme for enhancing the texture region. Here a local variance is calculated from neighbouring pixels to understand the texture degree of pixels. These pixels are sorted in descending order of the local variance. Smooth areas correspond to low local variance values. Further processing is done using a rhombus prediction pattern. Using the generated PEs via rhombus prediction, the cover image is divided into two parts which are independent sets named as dot and cross sets. Now, a two-layer approach is used for embedding, i.e., the dot set first predicts the cross set while embedding data into PEs, and then the modified cross set predicts the dot set and embeds data into PEs. In general, PE is calculated initially, enumerating PEs will generate the histogram. The first and last bins are selected as the minimum value bin, and bin at a length equal to the size of metadata respectively. Data is embedded into all bins using the histogram shifting method and based on the descending order of the local variance while keeping unused bins unchanged.

After carrying out a detailed literature survey, it could be observed that all of the works have focused on one or more of the following points:

- Enhancing embedding of metadata rate and extraction of the complete information at the receiver end.
- Original image recovery without any quality deterioration.
- Image contrast enhancement.
- Reducing the visual distortions introduced due to the embedding of the metadata.
- Tamper detection or integrity authentication of ROI capability.
- Due consideration to split ROI and RONI.

Table 1 shows a comparative chart of a few of the state-of-the-art schemes which are based on reversible watermarking and reversible data hiding in medical images.

3. Application Scenarios

This section details a few of the applications related to the RDH.

Clinical Data Transferring Along with the Medical Image

Hospital information systems have been developed to meet the requirements of new modes of information sharing and remote communication. One of the applications of RDH in the medical field is the secure communication of clinical information like EPR embedded within the medical image for a second opinion or a detailed diagnosis. The design shown in Fig. 2 embeds the health report within an encrypted version of the original medical image using an RDH scheme and transfers the resulting image to the receiver. This resulting image which is a health report embedded image is decrypted at the receiver side using the decryption key and further processed using the RDH scheme to recover the original medical image and extract all the information embedded in it.

Table 1. Details of a few existing schemes.

Scheme	Embedding region	Technique used	Embedding rate (bpp)	Quality parameters
[5]	Both ROI and RONI	Difference Expansion and DWT	0.46–0.5	Avg. PSNR = 77.6dB Avg. SSIM = 0.9526
[9]	ROI	LSB	0.89–0.39 for ROI size 5% to 30%	Avg. PSNR = 43.98dB Avg. MSSIM = 0.9904
[16]	ROI	LSB	0.5 for ROI size 5%	Avg. PSNR = 102.25dB for single LSB plane
[11]	Border pixels of RONI	LSB	Not presented	Avg. PSNR = 52.13dB Avg. MSSIM = 0.9246
[31]	RONI	DCT	>= 1.63	PSNR <= 48.53dB
[29]	Both ROI and RONI	Dual tree complex wavelet transform and LSB	0.04–0.5	Max PSNR = 107.6dB Avg SSIM = 0.999974 (at 10% RONI)
[38]	Both ROI and RONI	Histogram shifting	0.01–3	PSNR = [14.74–23.06]dB SSIM = [0.3504–0.9047]
[30]	Entire image	Integer wavelet transform and discrete gould transform	0.0625–0.25	PSNR = [60.42–65.23]dB SSIM = [0.9832–1) IF = [0.9178–0.9836]
[27]	ROIs smooth region	Difference Expansion	upto 0.25	PSNR = [70.63–99.94]dB SSIM = 1 IF = 1
[40]	RONI	DWT	1.13–5.09	Avg. PSNR = 56.22dB
[24]	Entire image	Histogram shifting	0.007–0.02	PSNR = [67.57–76.5]dB SSIM = 1
[12]	Both ROI and RONI	Histogram shifting and LSB	0.076–0.374	PSNR = [24.5–30.45]dB SSIM = [0.8761–0.9829]

Medical Image Authentication

Reversible watermarking is a method of hiding the watermark in a medical image. The scenario in Fig. 3 gives an application for authenticating the communicated medical image. Here the watermarked image is transferred to the receiver. A reversible watermarking scheme is applied at the receiver to extract the watermark to recover the original image completely. The protection on the image no longer exists after the removal of the watermark.

Judicial Evidence Preservation

RDH can be applied to preserve the image evidence taken from the crime investigation scene. While an investigation is being carried out, all the pieces

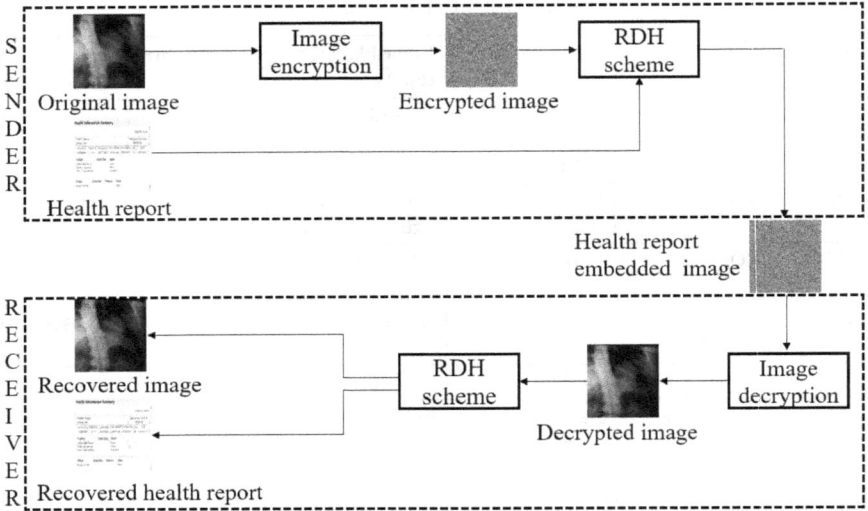

Figure 2. Health report transmission using RDH.

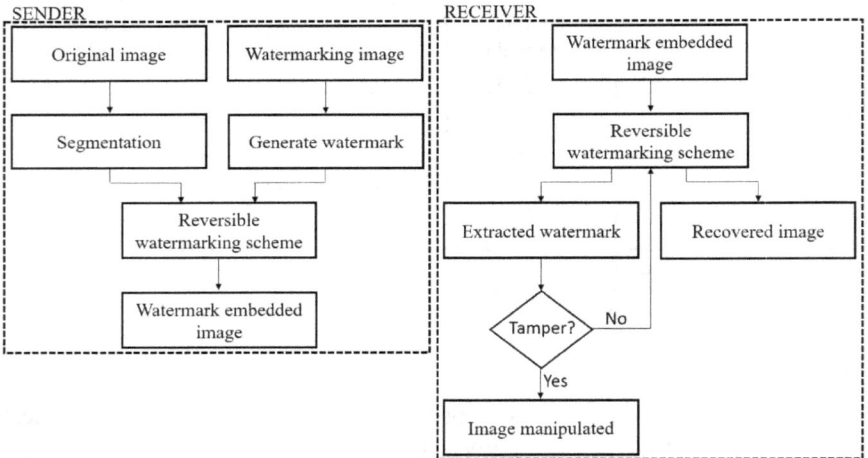

Figure 3. Medical image authentication using reversible watermarking.

of evidence gathered from the crime location need to be preserved safely so that the investigating team including law enforcement officers, prosecutors, judges, etc. can solve the crime and get justice for the victim. The team needs to initial, and mark the date and time along with the recording of other details accumulated from the spot, preserve, and distribute among officials. There is a high probability of culprits destroying the evidence or influencing the team. Hence, in order to secure the pieces of evidence, all the details of the scene can be hidden within the images taken from the location through RDH schemes.

Cloud Platform

Digital data including images has been growing every day and it is difficult to keep track of all the information by an individual. This information or data being generated can be stored and managed through a cloud facility. Everyone has started using the facility provided by the cloud service providers due to the convenience in accessibility and affordability. In the case of managing the images uploaded in the cloud, the cloud service providers need to maintain the details of the images via another file or record. Whenever a particular image is demanded by the user, the cloud service provider or administrator needs to access the file first to confirm the proper details and then the image needs to be retrieved and shared with the user. This is a cumbersome process. Hence, RDH can be adopted here to retain all the details of the image within itself. This will help in easing the maintenance and management of digital images in the cloud.

Image dataset

Medical images are required to carry out experiments and to evaluate the performance of different algorithms. Hence access to medical image databases is vital.

One of the open-access databases available for the research community is the Open Access Series of Imaging Studies (OASIS) [4], which consists of three OASIS brain data-set namely OASIS-1, OASIS-2, and OASIS-3. OASIS-1 comprises cross-sectional MRI data in young, middle-aged, nondemented, and demented older adults of 416 subjects aged 18 to 96 years. OASIS-2 comprises longitudinal MRI data in nondemented and demented older adults of 150 subjects aged 60 to 96 years. OASIS-3 is the latest introduction to the database which consists of longitudinal neuroimaging, clinical, and cognitive datasets for normal aging and Alzheimer's disease across 1098 subjects aged 42 to 95 years.

The National Biomedical Imaging Archive (NBIA) [3] is a web application containing a repository of in vivo images managed by the United States National Cancer Institute (NCI). NBIA provides access to cancer images to promote the requirements of research. Now, most of the cancer images have been migrated to The Cancer Imaging Archive (TCIA) [1], which is also available for open access. It contains medical images of different modalities like PET, CT, MRI, etc. stored in DICOM format along with a rich set of supporting data to help to supplement the investigation.

DICOM image sample data-sets library is exclusively available for research and teaching in [2]. OSIRIX was designed with the capability of viewing and traversing across multi-modality images as well as multi-dimensional images. The sample data-set includes medical images from CT, MRI, PET, MRA, XA angiogram, MR, CR, etc.

4. Challenges and the Future Scope of Research

- *Embedding capacity*: The embedding capacity indicates the quantity of information that can be embedded within the cover image. It is important to have a good embedding capacity to at least support the embedding of the required authentication information.

- *Maximum imperceptibility*: A good quality of the marked images needs to be ensured even after embedding the watermark or the metadata. The level of invisibility of the watermark reflects the imperceptibility factor.

- *Execution time*: Significant performance degradation of the hospital information management system should not happen due to the adopted data hiding scheme, i.e., the time taken for embedding the information at the sender side and extraction of information at the receiver side plays a significant role in overall execution time.

- *Robustness*: Any alterations to the medical image can change the cause of action after diagnosis. Hence it is recommended to detect the tamper effectively and to have a localisation function as well.

- *Lossless recovery*: The receiver should be able to without loss reconstruct the original image after removing the embedded watermark or metadata.

Evaluation Metrics

- *Bit error rate (BER)*: BER is the rate factor that indicates the amount of distortion introduced in the medical image. If I_X gives the count of incorrect bits in the extracted data and I_I is the total number of information originally embedded, then:

$$BER = \frac{I_X}{I_I}$$

It is not possible to tolerate any amount of distortion while dealing with medical images. Hence the value of BER should be 0.

- *Embedding capacity*: Embedding capacity is the measure of additional information (either watermark or metadata) which is carried over by the cover image. It is the quantity of additional information (bits) that a pixel of the medical image can hold. The unit of measurement is bits per pixel (bpp).

• *Peak signal-to-noise ratio (PSNR)*: The quality of the recovered or generated medical image can be measured using a metric such as PSNR. It is calculated with reference to a base image, given by the proportion of the peak signal strength to the noise factor in the signal. PSNR is computed by:

$$PSNR = 10log_{10}\left(\frac{(PEAK)^2}{MSE}\right)dB$$

where *PEAK* is the highest pixel intensity in the image and *MSE* is the noise measurement quantity called mean square error. Here the original cover image and the recovered image are compared to compute the *MSE*. The *MSE* equation is given by:

$$MSE = \left(\frac{1}{HW}\sum_{l=0}^{H-1}\sum_{m=0}^{W-1}(CI(l,m) - RI(l,m))^2\right)$$

where *H* indicates the height of the image matrix, *W* indicates the width of the image matrix, *CI* is the original cover image and *RI* is the recovered image.

• *Structural similarity index (SSIM)*: SSIM is a quality measuring metric based on human visual perception. It gives a measure of the structural degradation in the image being evaluated with reference to a base image considering the contrast, luminance, and structural components. If there is no degradation, then the value of SSIM is 1. In general, the value of SSIM is recorded between 0 and 1.

There are a few other parameters that can also be exploited by researchers for better quantifying their claims such as root mean square error (RMSE), the average difference (AD), mean absolute error (MSE), correlation measurement, image fidelity measure (IF), No Reference (NR) Image Quality Assessment (IQA) (NR-CDIQA), etc.

5. Conclusion

In this chapter, we have given a brief overview of data hiding in medical image communication. The main two areas of focus were reversible watermarking and reversible data hiding. A few of the existing research works in this domain was also discussed to give a glimpse of the research in the domain. A comparison table has been included which compares a few of the state-of-the-art schemes with specific parameters that help researchers with a quick reference. For a better insight into the applications of the techniques, a few scenarios are also disclosed. The challenges in carrying out the research work in the domain are specified, where the researchers could focus in the future to develop a better solution to the existing problems. A couple of medical images databases were also revealed that could be utilised in carrying out the experiments.

References

[1] The Cancer Imaging Archive (TCIA): Maintaining and Operating a Public Information Repository. Aug. 2021.

[2] DICOM Image Library - OsiriX. Aug. 2021.

[3] National Biomedical Imaging Archive – NBIA. Aug. 2021.

[4] OASIS Brains - Open Access Series of Imaging Studies. Aug. 2021.

[5] Al-Qershi, O.M. and Khoo, B.E. 2011. Authentication and data hiding using a hybrid roi-based watermarking scheme for dicom images. Journal of Digital Imaging 24(1): 114–125.

[6] Arsalan, M., Malik, S.A. and Khan, A. 2012. Intelligent reversible watermarking in integer wavelet domain for medical images. Journal of Systems and Software 85(4): 883–894.

[7] Balasamy, K. and Ramakrishnan, S. 2019. An intelligent reversible watermarking system for authenticating medical images using wavelet and pso. Cluster Computing 22(2): 4431–4442.

[8] Celik, M.U., Sharma, G., Tekalp, A.M. and Saber, E. 2005. Lossless generalized-lsb data embedding. IEEE Transactions on Image Processing 14(2): 253–266.

[9] Das, S. and Kundu, M.K. 2013. Effective management of medical information through roi-lossless fragile image watermarking technique. Computer Methods and Programs in Biomedicine 111(3): 662–675.

[10] Dragoi, I.C. and Coltuc, D. 2014. Local-prediction-based difference expansion reversible watermarking. IEEE Transactions on Image Processing 23(4): 1779–1790.

[11] Eswaraiah, R. and Reddy, E.S. 2014. Medical image watermarking technique for accurate tamper detection in roi and exact recovery of roi. International Journal of Telemedicine and Applications, 2014.

[12] Gao, G., Wan, X., Yao, S., Cui, Z., Zhou, C. and Sun, X. 2017. Reversible data hiding with contrast enhancement and tamper localization for medical images. Information Sciences 385: 250–265.

[13] He, W., Zhou, K., Cai, J., Wang, L. and Xiong, G. 2017. Reversible data hiding using multi-pass pixel value ordering and prediction-error expansion. Journal of Visual Communication and Image Representation 49: 351–360.

[14] Khan, S., Malik, S.A. et al. 2014. A high capacity reversible watermarking approach for authenticating images: exploiting down-sampling, histogram processing, and block selection. Information Sciences 256: 162–183.

[15] Lei, B., Tan, E.L., Chen, S., Ni, D., Wang, T. and Lei, H. 2014. Reversible watermarking scheme for medical image based on differential evolution. Expert Systems with Applications 41(7): 3178–3188.

[16] Liu, Y., Qu, X. and Xin, G. 2016. A roi-based reversible data hiding scheme in encrypted medical images. Journal of Visual Communication and Image Representation 39: 51–57.

[17] Makbol, N.M. and Khoo, B.E. 2014. A new robust and secure digital image watermarking scheme based on the integer wavelet transform and singular value decomposition. Digital Signal Processing 33: 134–147.

[18] Manikandan, V.M., Murthy, K.S.R., Siddineni, B., Victor, N., Maddikunta, P.K.R. and Hakak, S. 2022. A high-capacity reversible data-hiding scheme for medical image transmission using modified elias gamma encoding. Electronics 11(19): 3101.

[19] Memon, N.A. 2011. A novel reversible watermarking method based on adaptive thresholding and companding technique. International Journal of Computer and Information Engineering 5(7): 738–742.

[20] Natarajan, V. et al. 2016. Hybrid local prediction error-based difference expansion reversible watermarking for medical images. Computers & Electrical Engineering 53: 333–345.

[21] Nguyen, T.S., Chang, C.C. and Huynh, N.T. 2015. A novel reversible data hiding scheme based on difference-histogram modification and optimal emd algorithm. Journal of Visual Communication and Image Representation 33: 389–397.

[22] Ni, Z., Shi, Y.Q., Ansari, N. and Su, W. 2006. Reversible data hiding. IEEE Transactions on Circuits and systems for Video Technology 16(3): 354–362.

[23] Ou, B., Li, X., Zhao, Y., Ni, R. and Shi, Y.Q. 2013. Pair-wise prediction-error expansion for efficient reversible data hiding. IEEE Transactions on Image Processing 22(12): 5010–5021.

[24] Pan, W., Bouslimi, D., Karasad, M., Cozic, M. and Coatrieux, G. 2018. Imperceptible reversible watermarking of radiographic images based on quantum noise masking. Computer Methods and Programs in Biomedicine 160: 119–128.

[25] Pan, Z., Hu, S., Ma, X. and Wang, L. 2015. Reversible data hiding based on local histogram shifting with multilayer embedding. Journal of Visual Communication and Image Representation 31: 64–74.

[26] Peng, F., Li, X. and Yang, B. 2014. Improved pvo-based reversible data hiding. Digital Signal Processing 25: 255–265.

[27] Qasim, A.F., Aspin, R., Meziane, F. and Hogg, P. 2019. Roi-based reversible watermarking scheme for ensuring the integrity and authenticity of dicom mr images. Multimedia Tools and Applications 78(12): 16433–16463.

[28] Qin, C., Chang, C.C. and Liao, L.T. 2012. An adaptive prediction-error expansion oriented reversible information hiding scheme. Pattern Recognition Letters 33(16): 2166–2172.

[29] Ro˘cek, A., Slav´ı˘cek, K., Dost´al, O. and Javorn´ık, M. 2016. A new approach to fully-reversible watermarking in medical imaging with breakthrough visibility parameters. Biomedical Signal Processing and Control 29: 44–52.

[30] Selvam, P., Balachandran, S., Iyer, S.P. and Jayabal, R. 2017. Hybrid transform based reversible watermarking technique for medical images in telemedicine applications. Optik 145: 655.

[31] Shih, F.Y. and Zhong, X. 2016. High-capacity multiple regions of interest watermarking for medical images. Information Sciences 367: 648–659.

[32] Tai, W.L., Yeh, C.M. and Chang, C.C. 2009. Reversible data hiding based on histogram modification of pixel differences. IEEE Transactions on Circuits and systems for Video Technology 19(6): 906–910.

[33] Thabit. R. and Khoo, B.E. 2014. Robust reversible watermarking scheme using slantlet transform matrix. Journal of Systems and Software 88: 74–86.

[34] Tian, J. 2003. Reversible data embedding using a difference expansion. IEEE Transactions on Circuits and Systems for Video Technology 13(8): 890–896.

[35] Wu, H.T., Huang, J. and Shi, Y.Q. 2015. A reversible data hiding method with contrast enhancement for medical images. Journal of Visual Communication and Image Representation 31: 146–153.

[36] Xuan, G., Yang, C., Zhen, Y., Shi, Y.Q. and Ni, Z. 2004. Reversible data hiding using integer wavelet transform and companding technique. pp. 115–124. In International Workshop on Digital Watermarking, Springer.

[37] Yang, Y., Zhang, W., Liang, D. and Yu, N. 2016. Reversible data hiding in medical images with enhanced contrast in texture area. Digital Signal Processing 52: 13–24.

[38] Yang, Y., Zhang, W., Liang, D. and Yu, N. 2018. A roi-based high capacity reversible data hiding scheme with contrast enhancement for medical images. Multimedia Tools and Applications 77(14): 18043–18065.

[39] Yu¨zkollar, C. and Kocabıçak, Ü. 2015. Region based interpolation error expansion algorithm for reversible image watermarking. Applied Soft Computing 33: 127–135.

[40] Zhong, X. and Shih, F.Y. 2019. A high-capacity reversible watermarking scheme based on shape decomposition for medical images. International Journal of Pattern Recognition and Artificial Intelligence 33(01): 1950001.

[41] Zou, D., Shi, Y.Q., Ni, Z. and Su, W. 2006. A semi-fragile lossless digital watermarking scheme based on integer wavelet transform. IEEE Transactions on Circuits and Systems for Video Technology 16(10): 1294–1300.

CHAPTER 7

How Can We Enhance Data Security in Digital Twin-Enabled IoT Networks

A Proposal with ECAs Toward Prime Out-degree Finding of PageRank Regular Digraph in Blockchain

Apsareena Zulekha Shaik and *Arnab Mitra**

1. Introduction

Digital Twin technology is a promising concept that is gaining interest from practitioners and researchers from different backgrounds. This emerging technology suggests the ability to emulate any physical entity into its equivalent software (i.e., digital) existence. Thus, the role of Digital Twin might be realised in the coming days in technology-enabled developments, e.g., Industry 4.0. We find that Industry 4.0 represents a change in how manufacturing and the industry sector are approached. Smart manufacturing, commonly referred to as intelligent production, is the goal of Industry 4.0. But there is a distinction between the two. The fusion of manufacturing with Artificial Intelligence (AI) has given rise to intelligent manufacturing. Smart manufacturing is the integration of intelligent manufacturing with smart

Dept. of Computer Science & Engineering, SRM University-AP, Andhra Pradesh-522240, India.
Email: apsareenazulekha_shaik@srmap.edu.in
* Corresponding author: arnab.m@srmap.edu.in, mitra.arnab@gmail.com

technologies such as the Internet of Things (IoT), Cloud Computing (CC), Big Data Analytics (BDA), Cyber-Physical Systems (CPS), and Digital Twins (DTs) after the further improvement of AI to AI 2.0. As a result, the system becomes more intelligent, and effective, with the IoT and CPS both playing a significant role [1, 2]. To enhance the readability, a brief discussion of several of the latest technologies which are most frequently used is presented in the next chapter.

From our studies, we learned that a synergistic integration of several systems engineering concepts and control fields has been the main source of the CPS concept [3]. Further, we learned from [4] that "… CPS is a system of cooperating computing units managing physical entities…" [4]. To support the said cooperating computing, software-based network components are typically used to connect all electromechanical components. We observed that the physical implementation of CPS is typically supported by the incorporation of IoTs. IoT enables the gathering and exchange of data over the Internet [2]. Although we observed the dependency between CPS and IoT, still we are able to find a scope to distinguish between them. We find that CPS and IoT both aim to strengthen connections between cyberspace and the physical world. However, the IoT places more of an emphasis on networking, whereas CPS places an emphasis on information exchange and feedback collection. On the other hand, DT addresses the need for information transparency, which is a prerequisite to the gathering, managing, and organising of data from devices in the IoT [6]. As already introduced, we find that DT is a virtual representation of a process, an asset, or a system that contains characteristics and behaviours of that entity that are appropriate for sharing, storing, interpreting, or processing in a certain environment [6]. By modelling what-if scenarios and analysing how modifications to the physical system affect the performance, DT is responsible for better decision-making. It is able to perform predictive maintenance since it can spot possible problems and failures before they happen. For the purposes of performing its functionalities, DT should offer a shareable database. As a result, concerns about integrity and confidentiality are urgent, making security a crucial requirement. To support these, Blockchain Technology, which is typically a distributed ledger, offers a cutting-edge solution for sharing data among several parties that places an emphasis on security aspects like integrity [7]. We find that Blockchain technology might be used to store sensor data and make it instantly accessible. The counterpart or DT is then used to synchronise this data [7]. Since unchangeable data is stored involving several consensuses, e.g., Proof of Work, Proof of Stake, Delegated Proof of Stake, etc. [23], Blockchain Technology is often considered more secure as compared with other technologies, e.g., CC, BDA, AI, IoT, etc. [8]. Additionally, it provides a dispersed network and upholds system integrity, removing the need for centralised control. To enhance the existing Blockchains, several researchers have proposed possible enhancements. Among them, enhancements of Blockchains involving PageRank are particularly of interest. A cost-effective computation of reputation score was presented in [9] which is the extension of the PageRank algorithm and a novel consensus for Delegated Proof of Reputation (DPoR)-based Blockchain is designed. As our chapter is influenced by PageRank-based DPoR toward Blockchain, a brief discussion on PageRank is presented next.

One of the popular methods for determining a web page's rank is the use of the PageRank algorithm. The PageRank algorithm mostly uses a webpage's in-links and out-links to determine a rank [10]. We found that the PageRank computation may be achieved with the help of directed, weighted, temporal, multi-relationship blockchain transaction graphs [11] and or digraphs [12] to infer the relevance between accounts when searching graphs, those we find are typically based on the in-links and out-links of a node. The diagrammatic representation of nodes interconnected using in-links and out-links is presented in Fig. 1.

As digraphs play a major role in PageRank algorithms [12], a brief discussion on digraphs is presented next. We learned from [12] that digraphs are important in view of data privacy and security. Readers of this chapter may further go through [12] to have a detailed insight into Digraphs with Prime Outdegrees.

A digraph is composed of a finite non-empty set V called vertices and A which are ordered pairs known as arcs. The digraph with a set of vertices V, and a set of arcs A is represented as $D = (V, A)$.

If (a, b) is an arc, it is said that a and b are adjacent to each other. The set of vertices adjacent from a given vertex 'v' is represented as $N^+(v)$; to a given vertex 'v' is represented as $N^-(v)$. $d^+(v)$ is the notation to denote the outdegree of a given vertex and the indegree of a vertex 'v' is denoted with $d^-(v)$ [12].

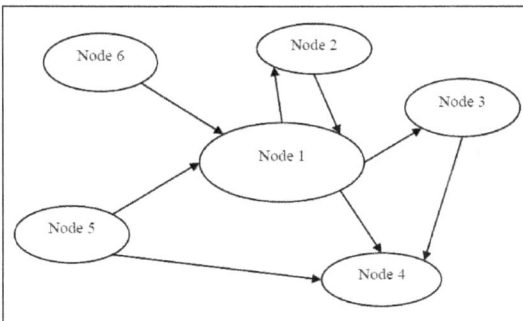

Figure 1. A typical interlinked structure for nodes in a network (as inspired by [19]).

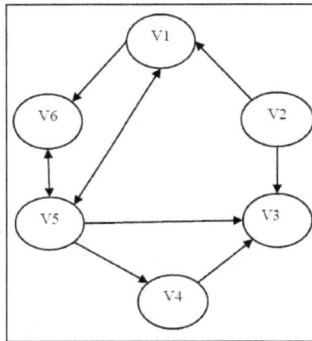

Figure 2. A typical diagram for a diagraph (as inspired from [12]).

One sample of digraph is illustrated in Fig. 2 which describes the similarity of Digraphs and connected websites (refer to Fig. 1).

Prime numbers and primality properties are very important in view of data security. For this reason, the inclusion of prime integers and or primality property may also be found are also added to the PageRank algorithms using digraphs. Both the outdegree sequence of PageRank uniform digraphs with prime outdegree and the degree sequence of PageRank uniform symmetric digraphs have been described [12]. For many scientific and engineering applications as well as security-related techniques, prime numbers are more crucial. There are various ways to determine whether a given integer is prime, but Fermat primality testing offers the highest level of certainty. Equations 1 and 2 provide evidence for Fermat's idea of primality [13]. Researchers in the past have investigated the use of Cellular Automata (CAs) in a prime generation [14–17]. Following Equation 1, and Equation 2 as studied from [24] once again presented those that describe the Fermat Primality checking algorithm.

$$A^P \equiv A(mod\ P) \tag{1}$$

where P is a prime; A is a natural number.

Furthermore, if $P * A$ (A is not divisible by P), then some minimum exponent P is found such that

$$A^{P-1} \equiv 1\ (mod\ P) \tag{2}$$

where, $A^{P-1} - 1$ is divisible by P.

On the other hand, we find that complex systems may be dynamically modelled using ECAs, which is a mathematical technique. The CA architectures have a number of dimensions. This chapter focuses on elementary cellular automata, which are 3-cell structures with two neighbours (ECAs). There are

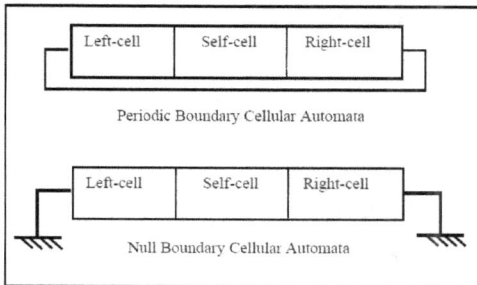

Figure 3. A typical diagram for ECAs at different boundaries (as inspired from [20]).

two situations involving boundaries. The first and last cells are connected in periodic boundary cellular automata (PBCAs); the first and last cells are grounded in null boundary cellular automata (NBCAs). The diagrammatic representation of PBCAs and NBCAs can be seen in Fig. 3.

As discussed in [18] there are few ECAs basics which are followed next:

- **Linear CAs rule:** If the next state computing function is defined using XOR or XNOR logic, then it is considered as a Linear CAs rule.

- **Non-linear CAs rule:** If the next state computing function is defined using AND, OR logic, then it is considered as a Non-linear CAs rule.

- **Homogeneous (uniform CAs):** If every cell of the CAs configuration is applied with the same CAs rule, then it is referred to as Homogeneous (uniform) CAs.

- **Heterogeneous (hybrid CAs):** If two or more cells of the CAs configuration are applied with different CAs rules then it is referred to as Heterogeneous (hybrid) CAs.

Uses of ECAs for several scientific and engineering applications may be found in a number of different kinds of literature. Among them, the usefulness of an ECAs-based PageRank validation methodology in the Cloud was discussed in [19] with reference to the physical components' low energy consumption. Furthermore, the suggested method by [19] guarantees improved performance for analytical and real-time internet traffic data analysis toward PageRank validation. It was further concluded in [19] that environmental sustainability depends on a system's stability and energy efficiency. As a result, a PageRank validation model based on energy stability was described in [20]. The experiments were conducted in [19, 20] to demonstrate that *MTCMOS* flip-flops have the lowest power consumption and $DT - TG$ flip-flops have the most. At 1.0 v_{dd} and 2.0 v_{dd}, it has been observed that each $DT - TG$ flip-flops require 1.20 $E - 05$ watts (W) and 1.20 $E - 0.5$ watts (W), respectively. Thus, it was concluded in [19, 20] that the ECAs-based model uses extremely low energy.

From our studies, though we have found several kinds of literature to upgrade the Blockchains, we still, did not find that a systematic study was ever presented to incorporate such modelling towards the construction of the energy-efficient simple construction and possible upgradation of existing Blockchain technology. For this reason, we find an opportunity to continue our research.

The major contributions of our presented chapter are as followed.

- It introduces a synergetic integration of several existing technologies, e.g., digraph, PageRank, and ECAs towards the possible enhancement of Blockchains.
- It further presents a novel design towards an enhanced Blockchain involving the prime outdegree finding of PageRank regular digraph for superior data security in digital twin-enabled IoT networks.
- An effective and easy implementation for the primality confirmation using Fermat's theorem using Python 3.10.6 is presented in this chapter to verify the prime outdegree associated with the Blockchain network which ensures the time complexity of $O(n * p^2)$.
- A python program using Python 3.10.6 is presented to find out the out-links for a node in a Blockchain network which ensures the time complexity of $O(n)$.

The architecture of the remaining chapter is as followed. Section 2 presents related works, Section 3 briefly describes the background works, Section 4 presents the proposed work; Section 5 presents the result analysis, and Section 6 presents the concluding remarks and future research direction.

2. Related Works

A detailed discussion on CPS and DTs was provided in [1]. Further, an analysis of CPS and IoT focusing on the similarities, and differences as well as the relationship between them was enlightened in [3]. In addition to this, definitions of the DT concepts were also derived specifically in view of Industry 4.0 in [2]. Along with diving into the definitions of the DT, all-inclusive definitions of IoTs: ranging from localised to global systems, were presented in [4] with sufficient insights towards several IoTs architectures and their specific requirements. Further, a future research direction, key challenges and scientific problems in view of IoTs were briefly presented in [5]. We found that DT and IoT standards were investigated in [6] to stimulate the consolidation of standards. Besides, the discussion of DT concerning challenges and solutions, a proper need was justified in [6] to address the security measures. Therefore, as a solution to secure DT's sharable database, the usage of novel technology of distributed ledgers was introduced in [7] that suppresses integrity and confidentiality issues.

Blockchain technology is the best example of a distributed ledger and hence we observe a rapid evolution of the Blockchain. The Blockchains and its importance in smart contracts was presented in [8]. Several attempts to enhance the Blockchains might be found in the literature. Among several others, we find, an enhanced consensus in Blockchain technology for Industry 4.0 was presented in [9]. We find that in [9], a proper justification was presented for the consideration of web PageRank in Blockchain networks toward the computation of reputation-based consensus. Thus, the PageRank algorithm was extended to provide this new consensus for DPoR (Delegated Proof of Reputation)-based Blockchain. The ECAs-based PageRank validation approach was utilised in [9] to accomplish this. The outdegree sequence of PageRank [10, 11] uniform digraphs [12] with prime outdegrees and characterisation of the degree sequence of PageRank uniform symmetric digraphs was discussed in [12]. The significance of PageRank uniform digraphs in the context of data security is discussed in [13]. An energy stability-aware strategy in ECAs-based PageRank validation model is proposed [18]. Researchers have placed a lot of emphasis on energy efficiency and low power usage in computers [19]. Thus, we observe the focus on energy-based PageRank computation and validation. The amount of content on a web page closely relates to its energy and page rank, as explained in [19].

In [18–21], several scenarios of PageRank, including stable traffic PageRank, increasing traffic PageRank, and decreasing traffic PageRank, were discussed. In [21], the RMT (Rule Mean Term) for the fundamental frame, increasing traffic, and decreasing traffic were considered. ECAs incorporation into the PageRank validation model added design simplicity and energy efficiency. An in-depth investigation of PageRank as well as the connection between page energy are presented in [21]. Additionally, a mathematical model of PageRank was discussed using discrete Green's functions [27] presented in [21] which ensured energy efficiency. Several other literature available [28–32] may also be observed, which further focused on the emphasis on PageRank and other ECAs-based applications. As our research in the presented chapter is greatly influenced by the works of [18–20, 22], we choose to present them briefly in Section 3.

3. Background

As studied in [20], we find that websites are indexed (i.e., ranked) mostly based on the number of in-links and out-links. To facilitate the ranking (i.e., indexing of websites) several static and dynamic PageRank computation algorithms have been presented by researchers. However, after ranking any website, we find an argue to validate the computed PageRank [18–20]. We found that in [20], using D flip-flops, ECAs were employed to accomplish page ranking validation at a low cost. Based on Langton's λ-parameter, ECAs rules at NBCA configuration were further investigated [20]. In addition to the significance of PageRank validation, a green strategy was also presented in [18–19] that preserved energy efficiency in the computed dynamic PageRank.

In [19], we observed a focus on environmental sustainability by ensuring energy efficiency in associated design. To ensure the system's possible competence in validating PageRank using ECAs at a low-cost physical implementation, a few experiments were conducted and it was concluded that a change in energy causes a change in the PageRank of a web page, i.e., ($\Delta E_{web\ page}$) α ($\Delta\ page\ rank$) where, $\Delta E_{web\ page}$ is referred to as change of energy for web page and $\Delta\ page\ rank$ is referred to as change of PageRank. It also examined the relationship between the energy changes and the amount of change in stored information and finally presented the relationship that ($\Delta I_{web\ page}$) α ($\Delta E_{web\ page}$) where, $\Delta I_{web\ page}$ referred to as a change of stored information in any web page.

The PageRank validation as discussed in [18–19] may be realised by the following set of equations Equation 3 [18], and Equation 4 [18].

$$E_{page\ rank\ validation} = E_{page\ energy different\ traffic} + \sum_{i=1}^{n} E_{physical\ component fabrication} \qquad (3)$$

$$E_{web\ page\ static traffic} = \sum \begin{bmatrix} E_{web\ page\ static traffic} \\ +E_{web\ page\ increased traffic} \\ +E_{web\ page\ decreased traffic} \end{bmatrix} \qquad (4)$$

Further, we observed from [18] that page traffic has a substantial influence towards the computation of PageRank. A web page's bandwidth is a crucial component of its energy stability [21]. The average page load time for a website significantly influences its PageRank trends in the event of constant bandwidth [18]. As a result, an index named *StabilityIndex$_{web\ page}$* has been employed to validate PageRank in terms of energy stability [18]. Equation 5 explains *Stability index$_{web\,page}$* of PageRank validation.

$$Stability\ Index_{web\,page} = \left(\begin{array}{c} percentage\ of\ change\ in\ page\ energy \\ \times \\ change\ in\ average\ load\ time \end{array} \right) \quad (5)$$

Another important aspect is data security where the application of Prime numbers may be observed. We found that CAs may be linked with the cost-efficient Primes finding method to improve data security and privacy. In [22], prime sequence generation using ECAs is provided as a flip-flop cost implementation of ECAs-based modelling. Here, the fixed boundary environment has been used to investigate the natural sequence of primes, or primes A000040, which improves data security and privacy. At an automaton with a size of 8, the first 50 numbers of the A000040 series can be explored. A hybrid ECAs produced by the ECAs rules 60 and 204 is discovered to produce every cycle in its transition diagram [22].

4. Proposed Work

Based on the ECAs-based design as briefly discussed in Section 3, we presented a synergetic integration of several ECAs-based models to enhance the Blockchains. Figure 4 presents an operational flowchart that illustrates how energy-based PageRank towards enhanced Blockchain consensus is considered. The computation of out-links of a node and primality test using Fermat's theorem as found in Fig. 5 and Fig. 8 that are described in algorithms 1, and 2. Firstly, the out-links of a node are calculated and it checks whether the number of out-links are prime or not using Fermat's theorem. This helps us to ensure that the node has a prime out-degree. Algorithm 1 was designed to compute the number of out-links of a node. Algorithm 2 checks the primality of the output we got from algorithm 1. Table 1 contains the results we got on calculating the out-links of each node.

Algorithm 1. Calculating the number of out-links in the given website

Input: URL for a website.
Output: Number of out links associated with that website.
1. Import BeautifulSoup and requests libraries
2. Set the URL of the web page to be scraped
3. Use the requests module to get the HTML content of the web page
   webpage = requests.get(url)
4. Extract the HTML content using the text attribute of the response object
   page_data = webpage.text

5. Create a BeautifulSoup object from the HTML content with the 'lxml' parser
   soup = BeautifulSoup(page_data, features = 'lxml')
6. Initialize a variable 'cnt' to 0
7. For each 'a' tag found on the web page:
   for i in soup.find_all('a')
   cnt = cnt+1 # Increment the 'cnt' variable
8. Print the total number of out-links found on the web page.

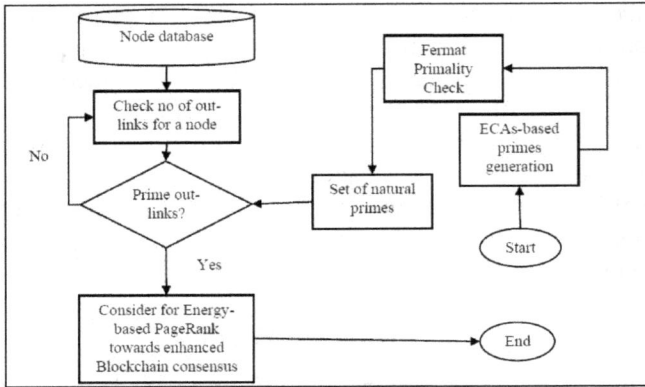

Figure 4. Proposed system flowchart.

```
1    from bs4 import BeautifulSoup
2    import requests
3
4    url = "https://flipkart.com"
5
6    web_page = requests.get(url)
7    page_data = web_page.text
8    soup = BeautifulSoup(page_data, features='lxml')
9    cnt = 0
10   for i in soup.find_all('a'):
11       link = i.get('href')
12       if link != "None" and link != "#" and link != url:
13           cnt+=1
14   print("No of outlinks in the website '"+url+"' are: "+str(cnt))
15
16
17
18
19
20
```

Requesting the page data

Iterating over the page data which has <a> tags

If a valid link is found in <a> tag 'cnt' is incremented

```
PROBLEMS   OUTPUT   DEBUG CONSOLE   TERMINAL                    Python + ∨ □

PS C:\Users\DELL> & C:/Users/DELL/AppData/Local/Programs/Python/Python310/python.
ter_8/book chapter/links_of_website.py"
No of outlinks in the website 'https://flipkart.com' are: 363
PS C:\Users\DELL>
```

Figure 5. A screenshot for the code generated involving Algorithm 1.

In Fig. 4, the nodes (typically URLs of a website as suggested by [9] instead of real nodes of a Blockchain network) are considered to examine the number of out-links of each node is calculated and on the other hand, the ECAs-based prime numbers are generated. We request readers to go through [22] for detailed information regarding ECAs-based prime generation. These generated out-link numbers are passed through the Fermat Primality test (refer to Algorithm 2) for verifying for the second time whether they are primes. The final set of primes after the Fermat Primality test are called primes. The number of out-links 'n' of each node is now checked whether the 'n' is part of the natural primes. If yes, that particular node is further considered for Energy-based PageRank towards enhanced Blockchain consensus.

Algorithm 2. Testing the primality of a number using Fermat's theorem

Input: a file with odd numbers.
Output: a file with prime numbers from the given input.
    # Define the gcd function using Euclid's algorithm
1. function gcd(a, b):
2. while b is not 0:
    temp = a
    a = b
    b = temp % b
3. return a
    # Define primalityTest function to check if a number is a prime
4. function primalityTest(p):
5.   for i from 1 to p-1:
6.   if gcd(p, i) is not 1:
        return 0
7.   else:
      if ($i^{\wedge}(p-1)$ mod p) is not 1:
        return 0
8.   return 1
    # Open input and output files
9. file = input from command line argument
10. inp = open(file, 'r')
11. lines = read all lines from inp
12. f = open("output.txt", "w")
    # Iterate through each line in the input file
13. for i in lines:
14.   if i is not equal to 1 and primalityTest(i) is true:
      f.write(i)
    # Close input and output files
15. inp.close()
16. f.close()

The above code is a Python program that reads a list of odd integers from a file and checks whether each integer is a prime number using the Fermat *primalityTest* function. If the integer is prime, it writes the number to a file called output.txt. The *primalityTest* function checks if a given integer p is prime or not by iterating over all possible values of a ($1 \leq a \leq p$), and checking if the conditions for primality are satisfied, namely if $\gcd(a, p) = 1$ and $a^{p-1} \equiv 1 \ (mod \ p)$. The gcd function is used to compute the greatest common divisor of two integers. The input and output images of Fermat's primality test can be seen in Fig. 6 and Fig. 7 respectively.

5. Related Analysis and Results

Figure 5 and Fig. 8 are the screenshots we took while the execution of the code using algorithms 1 and 2 in Section 4 respectively. Figure 6 and Fig. 7 are the screenshots of the input file and output file respectively that we got after executing the code using Algorithm 1. Table 1 presents the results on

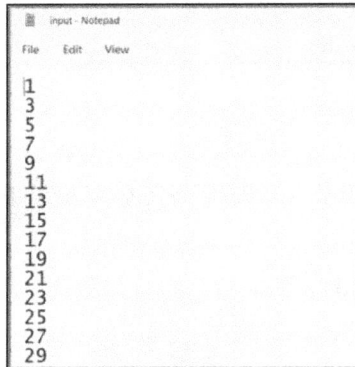

Figure 6. A screenshot from the input file used in Algorithm 1.

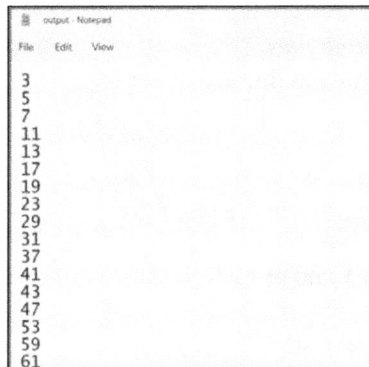

Figure 7. A screenshot for the output file used in Algorithm 1.

Table 1. The number of Out-links of nodes calculated using Algorithm 2.

Serial No	URLs	Number of Out-links	Serial No	URLs	Number of Out-links
1	https://www.google.co.in	27	26	https://www.mit.edu	59
2	https://www.yahoo.com	113	27	https://www.netflix.com	16
3	https://www.wikipedia.org	330	28	https://www.walmart.com	57
4	https://www.youtube.com	14	29	https://www.ebay.com	308
5	https://www.amazon.in	121	30	https://www.duckduckgo.com	2
6	https://www.flipkart.com	363	31	https://www.airbnb.com	3
7	https://www.whatsapp.com	69	32	https://www.reddit.com	143
8	https://www.github.com	107	33	https://www.ikea.com	132
9	https://www.vimeo.com	105	34	https://www.ndtv.com	420
10	https://www.linkedin.com	128	35	https://www.nytimes.com	295
11	https://www.apple.com	110	36	https://www.researchgate.net	32
12	https://www.wordpress.org	76	37	https://www.srmap.edu.in	197
13	https://www.indiatimes.com	161	38	https://www.cambridge.org	139
14	https://www.facebook.com	43	39	https://www.cnn.com	247
15	https://www.msn.com	132	40	https://www.sciencedirect.com	12
16	https://www.medium.com	106	41	https://www.merriam-webster.com	159
17	https://www.paypal.com	71	42	https://www.indiatoday.in	289
18	https://www.forbes.com	191	43	https://www.alibaba.com	137
19	https://www.bbc.com	306	44	https://www.slideshare.net	91
20	https://www.scribd.com	89	45	https://www.weebly.com	96

Table 1 contd. ...

...Table 1 contd.

Serial No	URLs	Number of Out-links	Serial No	URLs	Number of Out-links
21	https://www.businessinsider.com	364	46	https://www.indiamart.com	990
22	https://www.twitter.com	6	47	https://www.stackoverflow.com	111
23	https://www.bing.com	4	48	https://www.byjus.com	829
24	https://www.nytimes.com	299	49	https://www.harvard.edu	157
25	https://www.microsoft.com	123	50	https://www.espn.com	159

calculating the out-links of each node (i.e., home page of each web site as discussed in [9]).

Time Complexity of the Website Out-links Program

The time complexity of this code depends on the size of the web page being fetched and the number of hyperlinks in it. The *find_all()* method has a time complexity of $O(n)$ where 'n' is the number of elements searched for. Therefore, the overall time complexity of the code is $O(n)$, where 'n' is the number of hyperlinks in the webpage.

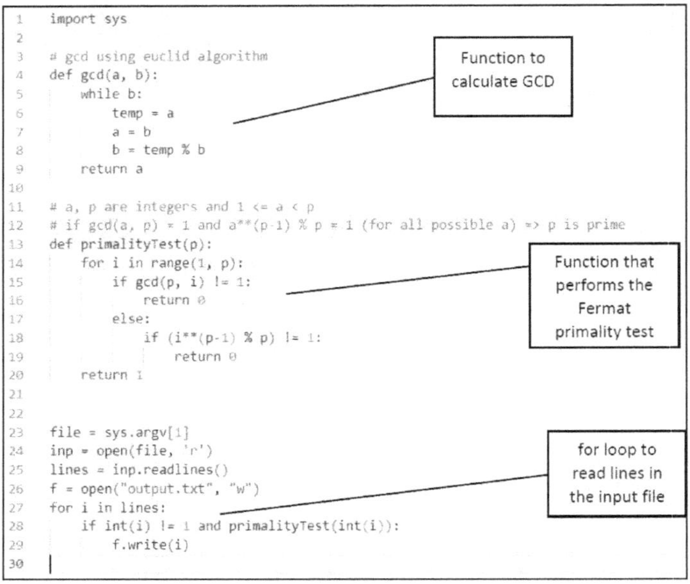

```
1   import sys
2
3   # gcd using euclid algorithm
4   def gcd(a, b):
5       while b:
6           temp = a
7           a = b
8           b = temp % b
9       return a
10
11  # a, p are integers and 1 <= a < p
12  # if gcd(a, p) = 1 and a**(p-1) % p = 1 (for all possible a) => p is prime
13  def primalityTest(p):
14      for i in range(1, p):
15          if gcd(p, i) != 1:
16              return 0
17          else:
18              if (i**(p-1) % p) != 1:
19                  return 0
20      return 1
21
22
23  file = sys.argv[1]
24  inp = open(file, 'r')
25  lines = inp.readlines()
26  f = open("output.txt", "w")
27  for i in lines:
28      if int(i) != 1 and primalityTest(int(i)):
29          f.write(i)
30
```

Function to calculate GCD

Function that performs the Fermat primality test

for loop to read lines in the input file

Figure 8. A screenshot for the code generated using Algorithm 2.

Time Complexity of Fermat Primality Test

The 'primalityTest' function has a time complexity of $O(p^2)$ where p is the input parameter. In the worst case, the range of the for loop in the 'primalityTest' function is from 1 to $p - 1$, so the time complexity of the function is $O(p^2)$. The time complexity of the '*gcd*' function is $O(\log(\min(a, b)))$ where a, b are the parameters. The code reads the input file, which takes $O(n)$ time where n is the number of lines in the input file. For each line, it checks if the number is not 1 and performs a primality test on the number, which takes $O(p^2)$ time. Therefore, the overall time complexity of the main function is $O(n * p^2)$.

Hardware Requirements

The hardware configuration used to perform the experiment consists of an Intel® Core™ i5-8250U CPU clocked at 1.60 GHz with a boost frequency of up to 1.80 GHz. The processor is a quad-core processor. The hardware has 8.00 GB of installed RAM and equipped with 64-bit operating system and an x64-based processor.

We extensively considered different web sites as a replacement of nodes as suggested in [9]. The experimental results as achieved with our code (refer to Fig. 5) is presented in following Table 1.

The results towards prime out-degree finding involving our code (refer to Fig. 8) for those data of Table 1 are presented in the following Table 2.

From Table 2, it is observed that a decision toward prime out-degrees may be achieved at ease with a time complexity of $O(n * p^2)$ and at a reduced searching space (50% reduced searching space [22]). To explain it further, a simple illustration is presented next.

Illustration 1.

From Table 2, we observed that for "https://www.airbnb.com" (serial 31 of Table 1) has an out-degree of 3 which is a prime outdegree at a time complexity $O(n)$ (refer Algorithm 1). The ECAs-based state time diagram as discussed in [22] is once again presented in the following Fig. 9.

As < 204, 60, 204 > at several
fixed boundary scenarios

From Fig. 9, we observed that all odd and even states have been separated within a few cycles. As, except for number 2, every prime number is an odd number, we may only focus on the cycles involving odd numbers. Thus overall, 50% of the states of Fig. 9 are required to consider only finding a prime number. Additionally, the transition diagram of Fig. 9 is independent of the boundary value conditions. Thus, it ensures cost-effectiveness. For a

Table 2. Prime out degree finding for the out-links of Table 1.

Serial No	Number of Out-links	Prime out-degree	Serial No	Number of Out-links	Prime out-degree
1	27	No	26	59	**Yes**
2	113	**Yes**	27	16	No
3	330	No	28	57	No
4	14	No	29	308	No
5	121	No	30	2	**Yes**
6	363	No	31	3	**Yes**
7	69	No	32	143	No
8	107	**Yes**	33	132	No
9	105	No	34	420	No
10	128	No	35	295	No
11	110	No	36	32	No
12	76	No	37	197	**Yes**
13	161	No	38	139	**Yes**
14	43	**Yes**	39	247	No
15	132	No	40	12	No
16	106	No	41	159	No
17	71	**Yes**	42	289	No
18	191	**Yes**	43	137	**Yes**
19	306	No	44	91	No
20	89	**Yes**	45	96	No
21	364	No	46	990	No
22	6	No	47	111	No
23	4	No	48	829	**Yes**
24	299	No	49	157	**Yes**
25	123	No	50	159	No

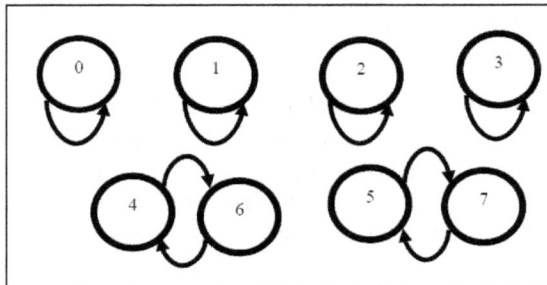

Figure 9. Transition diagram for ECAs at a heterogeneous EC.

more detailed insight into cost-effectiveness, we request the readers to go through [22].

6. Conclusions

A novel approach with ECAs is presented in this chapter towards enhanced data security and privacy in Digital Twin-enabled IoT networks. A combination of ECAs-based two approaches targeting PageRank validation and prime finding are incorporated in the presented approach to strengthen Blockchains by ensuring prime out-degree in regular digraphs for Blockchain nodes at a simple and cost-effective implementation. Thus, a cost-effective primality finding in PageRank ensures "...privacy preserving release of digraph data in environments where a dishonest analyst may have previous structural knowledge about the PageRank score..." [12]. Additionally, the proposed ECAs-based approach is found to be an energy-efficient model [18, 19, 22] which ensures its application suitability for energy and resource-constrained IoTs. Additionally, ECAs-based design supports inherent parallel computing and capability towards chip fabrication [9, 16] which might be beneficial in IoTs-based applications.

To conclude this paper, we would like to emphasise that the presented chapter is confined to presenting a theoretical basis towards enhanced Blockchains in an IoT-based model towards Digital Twins. We plan to carry out a detailed experimental implementation with supporting trials in near future toward actual use cases.

Declaration of Interests

The authors are not aware of the existence of any known and or, unknown conflicts. Hence authors declare no conflicts.

Acknowledgments

No research funding was availed to carry out the research and write the report. The authors further sincerely acknowledge that the reviews received from the anonymous reviewers have significantly helped to prepare the final chapter.

References

[1] Tao, F., Qi, Q. Wang, L. and Nee. A.Y.C. 2019. Digital twins and cyber–physical systems toward smart manufacturing and industry 4.0: Correlation and comparison. Engineering 5(4): 653–661.

[2] Negri, E., Fumagalli, L. and Macchi, M. 2017. A review of the roles of digital twin in CPS-based production systems. Procedia Manufacturing 11: 939–948.

[3] Greer, C., Burns, M., Wollman, D. and Griffor, E. 2019. Cyber-physical systems and internet of things.

[4] Minerva, R., Biru, A. and Rotondi, D. 2015. Towards a definition of the Internet of Things (IoT). IEEE Internet Initiative 1(1): 1–86.

[5] Ma, H.D. 2011. Internet of things: Objectives and scientific challenges. Journal of Computer Science and Technology 26(6): 919.

[6] Jacoby, M. and Usländer, T. 2020. Digital twin and internet of things—Current standards landscape. Applied Sciences 10(18): 6519.

[7] Dietz, M., Putz, B. and Pernul, G. 2019. A distributed ledger approach to digital twin secure data sharing. pp. 281–300. In Data and Applications Security and Privacy XXXIII: 33rd Annual IFIP WG 11.3 Conference, DBSec 2019, Charleston, SC, USA, July 15–17, 2019, Proceedings 33. Springer International Publishing.

[8] Hamilton, M. 2020. Blockchain distributed ledger technology: An introduction and focus on smart contracts. Journal of Corporate Accounting & Finance 31(2): 7–12.

[9] Mitra, A. 2022. How can we enhance reputation in Blockchain consensus for Industry 4.0–A proposed approach by extending the PageRank algorithm. International Journal of Information Management Data Insights 2(2): 100138.

[10] Sharma, P.S., Yadav, D. and Garg, P. 2020. A systematic review on page ranking algorithms. International Journal of Information Technology 12: 329–337.

[11] Wu, Z., Liu, J. Wu, J. and Zheng, Z. 2022. Transaction Tracking on Blockchain Trading Systems using Personalized PageRank. arXiv preprint arXiv:2201.05757.

[12] López, N. and Sebé, F. 2013. Degree sequences of pagerank uniform graphs and digraphs with prime outdegrees. pp. 303–313. In Combinatorial Algorithms: 24th International Workshop, IWOCA 2013, Rouen, France, July 10–12, 2013, Revised Selected Papers 24. Springer Berlin Heidelberg.

[13] Mitra, A. and Kundu, A. 2015, October. Analysis of sequences generated by ELCA-type cellular automata targeting noise generation. pp. 883–888. In 2015 19th International Conference on System Theory, Control and Computing (ICSTCC). IEEE.

[14] Wolfram, S. 1983. Cellular automata. Los Alamos Science, (Fall), 4–21. Retrieved from http://www.stephenwolfram.com/publications/academic/cellular-automata.pdf.

[15] Wolfram, S. 1984. Algebraic properties of cellular automata. pp. 71–113. Retrieved from http://www.stephenwolfram.com/publications/cellular-automata-complexity/pdfs/algebraic-properties-cellular-automata.pdf.

[16] Das, S., Kundu, A., Sikdar, B.K. and Chaudhuri, P.P. 2005. Design of nonlinear CA based TPG without prohibited pattern set in linear time. Journal of Electronic Testing 21: 95–107.

[17] Mazoyer, J. and Terrier, V. 1999. Signals in one-dimensional cellular automata. Theoretical Computer Science 217(1): 53–80.

[18] Mitra, A. 2019. On investigating energy stability for cellular automata based pagerank validation model in green cloud. International Journal of Cloud Applications and Computing (IJCAC) 9(4): 66–85.

[19] Mitra, A. and Kundu, A. 2017. Energy efficient CA based page rank validation model: a green approach in cloud. International Journal of Green Computing (IJGC) 8(2): 59–76.

[20] Mitra, A. and Kundu, A. 2015. Page ranking validation using cellular automata in cloud. International Journal of Cloud Applications and Computing (IJCAC) 5(3): 1–19.

[21] Guha, S.K., Kundu, A. and Dattagupta, R. 2016. Energy stability in cloud for web page ranking. pp. 621–630. In Proceeding of the International Conference on Information Systems Design and Intelligent Applications. Springer India. doi:10.1007/978-81-322-2752-6_61.

[22] Mitra, A. 2021. On the exploration of the natural sequence of primes with cellular automata targeting enhanced data security and privacy. International Journal of Cognitive Informatics and Natural Intelligence (IJCINI) 15(4): 1–18.

[23] Monrat, A.A., Schelén, O. and Andersson, K. 2019. A survey of blockchain from the perspectives of applications, challenges, and opportunities. IEEE Access 7: 117134–117151.

[24] Mitra, A. and Kundu, A. 2013. Cost optimized set of primes generation with cellular automata for stress testing in distributed computing. Procedia Technology 10: 365–372.

[25] Sampath, P. and Ramya, D. 2013. Performance analysis of web page prediction with Markov model, association rule mining (ARM) and association rule mining with statistical features (Arm-Sf). IOSR Journal of Computer Engineering 8(5): 70–74.

[26] Berl, A., Gelenbe, E., Di Girolamo, M., Giuliani, G., De Meer, H., Dang, M.Q. and Pentikousis, K. 2010. Energy-efficient cloud computing. The Computer Journal 53(7): 1045–1051.

[27] Chung, F. 2010. PageRank as a discrete Green's function. Geometry and Analysis I ALM 17: 285–302.

[28] Bianchini, M., Gori, M. and Scarselli, F. 2005. Inside pagerank. ACM Transactions on Internet Technology (TOIT) 5(1): 92–128.

[29] Richardson, M., Prakash, A. and Brill, E. 2006, May. Beyond PageRank: machine learning for static ranking. pp. 707–715. In Proceedings of the 15th International Conference on World Wide Web.

[30] Kundu, A., Dutta, R. and Mukhopadhyay, D. 2006, December. An alternate way to rank hyper-linked web-pages. pp. 297–298. In 9th International Conference on Information Technology (ICIT'06). IEEE.

[31] Kundu, A., Dutta, R. and Mukhopadhyay, D. 2007, November. Converging cellular automata techniques with web search methods to offer a new way to rank hyper-linked web-pages. pp. 291–295. In 2007 International Symposium on Information Technology Convergence (ISITC 2007). IEEE.

[32] Agrawal, P. and Rao, S. 2012, March. Energy-aware scheduling of distributed systems using cellular automata. pp. 1–6. In 2012 IEEE International Systems Conference SysCon 2012. IEEE.

CHAPTER **8**

A Secure Sensor Based Soil Nutrient Management System for Better Decision Making in Smart Agriculture

Usha Sri Peddibhotla

1. Introduction

All across the world, soil is a crucial component of agriculture. The health of the soil and the plants growing within is significantly influenced by soil microbes. The intricacy of preserving healthy soil is increased by the soil bacteria, elevation, slope, climate, and other elements. Maintaining the nutrients in the soil increases the health of the crops by ensuring the health of the soil. When soil continues to function as a vital living ecosystem that supports humans, animals, and plants, it is said to be healthy soil. The health of the soil is impacted by excessive fertiliser use and ongoing irrigation techniques on the same piece of land over time. As a result, the soil's fertility began to decline due to the health issues caused by over-fertilisation and irrigation. As a result, managing soil nutrients is an important part of the agriculture. As a source of plant nutrients, commercial fertilisers, manure, and organic by-products are applied as a part of soil nutrient management to agricultural landscapes. It places an emphasis on the next two possibilities that is, improving soil health and supplying crops with the nutrients they need.

SRM University-AP, Andhra Pradesh, India.
Email: Ushasri_p@srmap.edu.in

In light of the aforementioned possibilities, soil nutrient management can be carried out in one of two ways:

I) Real-time monitoring of the soil's variables

II) Managing soil toxicity

I. Real-time Monitoring

Smart agriculture is created as a result of the incorporation of sensing, actuating, and computing technology into conventional agriculture. Ploughing, sowing, irrigation, fertilizers/pesticides, crop harvesting, and live animals are some of the steps that make up the smart agriculture system [1]. Most traditional farming issues, such as drought response, crop optimization, environmental concerns, pest control, over/under irrigation, and so on, can be resolved by integrating IoT, Agbots, drones, remote sensing, and AI into the various stages of smart agriculture.

Before planting the seeds of the intended crops and plants, the soil must be thoroughly examined and prepared. In order to determine the soil's nutrient status and in order to implement numerous key decisions at various stages of soil ploughing and crop management, soil analysis and examination are crucial.

Sensing Technologies for Real-Time Monitoring of Soil Health

Soil sensors can keep an eye on the health of the soil. Instruments called soil sensors are used to keep track of soil moisture levels. They can be divided into sensors that keep tabs on pH, salinity, nutrition, conductivity, moisture, and temperature. Soil Sensors refer to all of these several instrument types as a whole.

Soil sensors measure the following variables

- Soil moisture
- Soil temperature
- Soil pH
- Soil NPK
- Soil Salinity
- Soil toxicity

Soil moisture: Photosynthesis may be negatively impacted by a low-moisture soil. Since soil pores, contain both water and air, they are constrained by prolonged water scarcity, crops may wither and die as a result. Lack of moisture prevents plant roots from breathing, which causes rotting, which under extreme circumstances might cause plant death.

Soil moisture sensors can be used to keep track of soil moisture. By analysing the data, prompt decisions about irrigation requirements may be made, ensuring that the soil has the right amount of moisture for crop growth.

Soil pH: Each plant has a preferred soil pH level, and most plants have difficulty developing normally when the pH is either too high or too low. For the growth of crops and to lessen the occurrence of pests and diseases, it is advantageous to use a soil pH sensor to detect the acidity or alkalinity of the soil. Testing the pH of the soil is therefore crucial for agricultural purposes.

Soil NPK: Because it is the main component of proteins, nitrogen is essential for the growth of fruit, leaves, and stems. In order to give plants, the nutrients they need and to encourage plant growth, soil NPK sensors can identify nitrogen, phosphorus, potassium, and other elements in the soil.

Soil salinity: The quantity of salinity in the soil can either encourage or restrict plant growth at varying rates, depending on the type of soil and the species of plant. The salinity of the soil can be determined using soil conductivity sensors.

The measurements of the data can be captured via sensors and follow the procedure below for the transmission and analysis of the data.

Data collection: Soil sensors often have a sensing component that monitors a particular aspect of the soil, such as its temperature, pH level, electric conductivity, or moisture content.

Signal conversion: The sensing element transforms the observed data into electric signals, which are often converted to digital representation by the sensor's internal circuitry for simpler transmission and processing.

Signal Transmission: Using cable or wireless communication techniques, digital signals are sent to a computer or controller.

Data processing: The computer or controller analyses and processes the data after receiving the digital signals in order to extract relevant information, such as the pH value, soil moisture, and temperature.

Device Control: The controller can automatically control associated equipment, such as irrigation systems and weather stations, based on the results of the analysis, enabling automated management.

Let us understand the flow through which the sensors are connected and how the data is captured using sensors.

Wireless sensors gather precise root zone data from below ground as shown in Fig. 1, where real growth occurs, and send the data to the monitoring service, which transforms the data into useful information for the farmers.

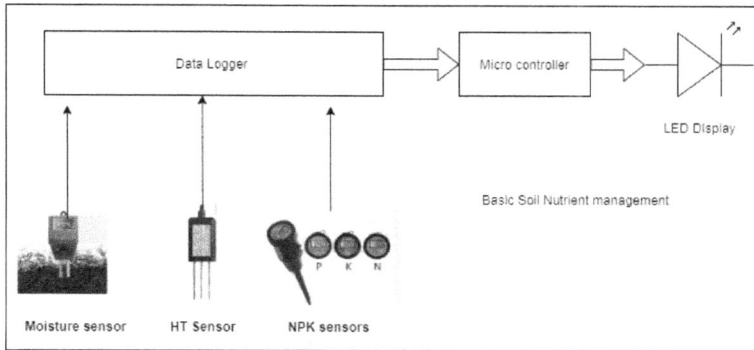

Figure 1. Basic soil nutrient management.

With this information, the farmers can now address their significant challenges and improve crop productivity and quality while lowering operational costs and water consumption.

A lot of agricultural information needs to be gathered district-by-district in order to create agricultural policies. Given the geographical diversity of India, where data must be separated and safeguarded for each state, the present security measures are insufficient to prevent data breaches and improper handling. The answer to it will be covered in the next section.

II. Managing Soil Toxicity

For agriculture to produce enough food to feed the globe, productive soils are required. The building of persistent poisonous substances, chemicals, salts, radioactive materials, and disease-causing organisms in soils, which have a negative impact on plant and animal health, is referred to as soil pollution. There are several ways by which soil can be poisoned, including seepage from a landfill, industrial release into the environment, release of contaminated water into the environment, and overuse of pesticides.

Industrial effluent discharge into neighbouring wetlands and farmland fields is the main contaminant affecting the ecosystem. The sort of gases or liquids emitted, the type of soil, and the vegetation of the land that is irrigated all affect how industrial effluents are applied. Polluted soil can change physically and chemically during irrigation, reducing the amount of nutrients available to both the crop and the soil [2]. Any layer of the soil can become polluted, including the surface and subsurface layers, which is then classified as agricultural soil pollution. Urban activities also contribute to soil pollution.

The subterranean channels can identify the levels of pollutants by using IOT sensors. The first stage in implementing a control measure is recognising the pollutant. The most common indicator for determining soil toxicity is the overall concentration of hazardous metals [4]. Due to the discharge of numerous industrial effluents, chemicals like mercury and carbon compounds are ingested by the soil. To detect soil pollution, sensors such as biosensors and soil NPK sensors are utilised, and the results are recorded in a data logger. This information will be displayed in the user interface for the activities related to agribusiness [5].

The introduction of IOT in the field of agriculture brought some challenges with it in terms of large data captured through real-time monitoring such as the following.

2. Security Issues

- **IoAT devices security:** There are estimated to be 40 billion connected IoT devices. The IoT devices are resistant to a variety of threats. However, the drive for inexpensive, basic hardware jeopardizes hardware security. The most frequent hardware security risks for IoT devices are hardware Trojan and side-channel attacks, which restrict the widespread use of IoT networks in crucial applications. Hardware Trojans use adversary-made modifications to the hardware that are hostile and can be used as a backdoor to take over the machine and launch attacks. These are extremely difficult to identify, and some techniques include using electronic microscopic scanning on de-metallized chips, examining power delays in the circuit, and looking at the PUF, which serves as an electronic device's distinctive signature. Another prominent hardware security risk is side-channel attacks, which exploit side-channel signals to retrieve private information like cryptographic keys. These problems are more common in IoT networks, and several fixes have been suggested [6, 7].

- **Data privacy and security:** Strong encryption techniques and security measures are required to guarantee data privacy and security during data transfers. The IoAT sensor nodes' constrained design and supporting protocols make it challenging to implement security measures using current technologies. Therefore, protecting user privacy and data is a major concern for smart agriculture.

- **Big Data challenge:** The sensors, actuators, and other IoAT devices will capture a substantial amount of random data. The usual approaches to processing this data are insufficient. It reduces concerns about food security and enhances the effectiveness of the entire supply chain in intelligent agricultural systems, it presents innovative business models, real-time decision-making, and predictive analysis [8]. Big data platforms for milk supply chain security have been integrated using support vector machines (SVM) and artificial neural networks (ANN). Data collection at various sensor nodes is the first step, and various data analysis techniques, including both conventional and big data analysis, are the last.

- **Scalability and reliability:** Depending on the size of the agricultural farm, different types of field sensors are needed. These sensors generate a varied amount of data which stresses the need for scalability in agriculture technology [5, 6, 11]. Cost will be reduced when the devices are reliable so that the number of redundant devices to accommodate fault tolerance.

- **Integration of Blockchain and IoT:** To address the above-mentioned security challenges, The use of blockchain in the IoAT looks to be quite pertinent. Data security and privacy are critical aspects of smart agriculture that must be addressed for autonomous processes to function well. The blockchain processes transactions in reverse chronological order and makes use of cryptographic safeguards to preserve data integrity and guard against adversary attacks like Denial-of-Service (DoS) and False Data Injection. Other problems with data, besides privacy, include ownership and monetization. In contrast to centralised applications where the data is monetized by a central authority, blockchain-based applications can assist farmers in managing the data access at granular levels and can help in monetizing the data on their own. In a typical IoT design, the edge layer's data is processed and stored in the cloud layer in order to carry out automated operations. The fact that the latency and access times in such networks fluctuate depending on network availability and the volume of access requests being sent to the server at any given time is one of their major downsides. Because decision-making requires real-time operations, the blockchain can aid in the creation of an effective real-time data-sharing paradigm. Some secure blockchain approaches for safe data sharing are put forth in [18, 19, 20, 21].

The schematic representation of a technological model introduces a thorough approach which solves various security problems from different perspectives. This model guarantees better soil nutrient management in terms of monitoring the soil nutrients in real time and reducing soil toxicity by alerting the concerned authorities.

The participants included in this model are as follows:

- Farmers who have agronomic knowledge and update the data to the higher levels from the sensors.
- Farming community authorities who collect data, suggest and decide the necessary technologies that are relevant to the problems raised.
- Other actors include the factory owners around the agricultural land, and agricultural authorities which also contribute to the services designed in the model.

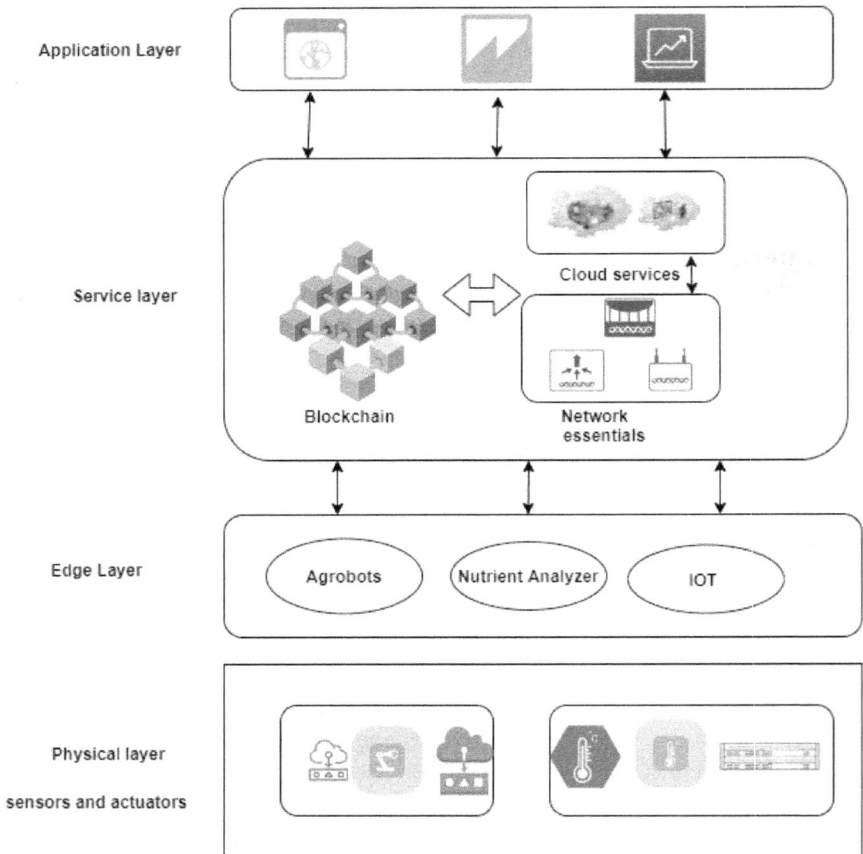

Figure 2. Layer-wise IoT Blockchain platform.

Our planned architecture has the layered architecture in Fig. 2, in which the services and functionalities of the proposed model are combined.

In the **physical layer or perception layer** IoAT devices are installed, which enable technicians or farmers to interact by entering information and requests. The processing layer communicates with the higher layer (IoT), filters data, and carries out control operations.

In **Edge Layer,** To handle the various operations in the architecture, information will be received from the higher layer. The IoAT instalment can differ depending on each design depending on the industry, type of cultivation, or environmental considerations. It must provide the help necessary for the data to be communicated to the installed blockchain in any case. Data immutability which can be gathered through secure IoT communication protocols is the goal of the combined use of blockchain and IoT.

The IoT solution installed in each facility is connected to a blockchain that is integrated into the **service layer**. This layer is in charge of managing different local and fog networks, communicating data from each facility, and sending and receiving data to and from various top-layer applications.

In the **Application layer,** applications are designed as per the scope of the usage such as desktop, mobile, and office networks, through the use of chain codes to interact with the blockchain which is the link of each application with the blockchain.

During integration of IoT with Blockchain, the following are the key functionalities that the IoT must perform:

- Capturing the data from the sensors.
- Installation of actuators for the purpose of human or actuator control using the User interfaces.
- Develop user-friendly mobile and web interfaces so that the events are recorded accurately.
- Deploying the data networks both locally and enabling sharing via the cloud and connected to the blockchain.
- Providing proper documentation and operating manuals.

Blockchain networks improve IoT facilities by enhancing their features. They offer unchangeable digital records, data veracity, and the capability to create algorithms that establish relationship models. These smart contracts are algorithms that offer a highly potent tool to produce the aforementioned advantages. In terms of IoT, the facility can modify the blockchain's design in some way. Nevertheless, it must offer the next features in each installation:

- Without relying on outside parties, verification is possible with blockchain technology.
- A blockchain's data structure is appendable only. Since the transactions recorded in each block are hashed in a way that cannot be changed or deleted, the data cannot be edited or destroyed.

After the process of verification, all transactions and data are attached to a block and transactions are stored in chronological sequence. As a result, the blockchain's blocks are all time-stamped.

- It is dispersed throughout all of the blockchain's active nodes.
- The transactions included in the blocks are spread across several computers involved in the chain. It is decentralised as a result. Additionally, it guarantees that any deleted data can be restored.

- It lessens the chance of fraud or double entry.
- Companies can establish conditions using smart contracts. Only when software algorithms are used to create the circumstances, are automatic transactions activated.

The coordination of the many players' actions and, on the other hand, the fundamental building blocks that the blockchain creates in accordance with the aforementioned functionality. It provides an overview of how the planned IoT Blockchain platform operates and how each component works. For using services like device and user registration, The app's client offers a platform through which users can propose transactions to the blockchain network. Registration with a certificate containing the private keys to sign the transaction is necessary before submitting it. A transaction is the act of reading or writing data from the blockchain network. Different devices and authorised users (farmers and technicians) can submit a transaction to register a new device or create a new task. The operation request is sent to the blockchain network by the IoT facility. It can also send the client's request to the device and receive status updates or sensing data from the device. The identification of the device's owner is validated, and transactions can be submitted directly from that person's smartphone. The status or sensor data is subsequently added to the ledger and compared to the smart contract parameters. A notification will be generated if the values recorded exceed specific thresholds.

Apps and communication interfaces allow device users (IoT, farmers, and others) to manage physical equipment without having any prior knowledge of them. Smart contracts are used to host the ledger functionalities over the network and allow controlled access to the data (Fig. 3). We also define an access control policy in the proposed platform, which permits users to access

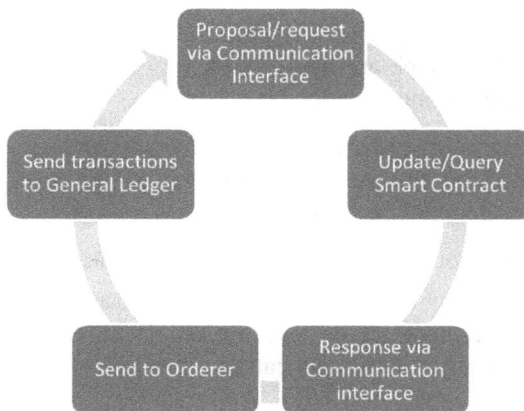

Figure 3. Workflow of a transaction.

a set number of approved contents or transactions. For instance, access to and control of the gadget are only allowed by the owner. Data storage methods are utilised to handle a huge amount of IoT data because Hyperledger is not designed for large transaction data payloads. A unique database (DB) is used in the system. To support huge file storage and reduce duplication throughout the whole blockchain filesystem, this DB is present on each peer. The blockchain must be configured, as well as the appropriate data transmission procedures, user apps, and database structures. The model and organisational rules between participants are included in the setup of the blockchain and depend on the installation type and predetermined requirements.

The first step in creating a Hyperledger network structure for one's application is listing the organisations of the participants. A security domain and a unit of identity and credentials exist within an organisation. It governs one or more network peers and is reliant on a membership service provider to issue identities and certificates to peers and clients for smart contract access privileges. The ordering network, which serves as the foundation of a Hyperledger network, is usually assigned its own organisation network. The sample network will consist of the following organizations: Farming community network, soil agriculture authorities, Industry owners and a state agency. The farming community periodically monitors the soil condition and updates the values to soil agriculture authorities. The authorities receive the data in the form of sensory signals and signals are transmitted to the edge layer.

To maximise security and expenses, various entities are combined into a single entity. In the capacity of a client, an entity receives permission from its organisations to submit transactions or read the ledger state. As a result, the blockchain network needs numerous peers, each of which represents a separate company. There is also an ordering service, one MSP for each of the four enterprises, and the peers on the network. A Hyperledger ordering node that accepts order transactions implements the ordering service. Hyperledger ensures the accuracy of every block that has been verified by a peer using deterministic consensus techniques.

An alert message is delivered to the industry authorities, the farming community, and the irrigation authorities via the user interface whenever a query is uploaded for the purpose of checking the threshold values.

3. Challenges & Future Scope of Research

The basic data characterisation of the soil can be used to determine the soil parameters at each field, such as the presence of rock fragments, bulk density, moisture, water content, carbon, salt, pH, carbonates, phosphorus, clay, sand, and silt mineralogy. Even though the blockchain offers a wide range of potential uses in smart agriculture to improve data security and integrity, there are still issues that must be resolved before this technology is widely adopted in the agriculture sector. The blockchain's consensus mechanism and cryptography components demand a lot of power and computing, in contrast to IoT technology, which is resource-constrained in terms of both power and computation. Research is being conducted to develop a variety of effective consensus methods that can be used in these contexts with limited resources, such as in smart agriculture. Another key issue that must be solved for widespread adoption is data. Large amounts of data like photos cannot be kept on the blockchain since each block's size is predetermined and constrained. In order to ensure secure access and data integrity, many researchers are striving to store data off-chain while also storing transaction and access metadata alongside the data on-chain.

Conclusion

The following findings from the analysis, design, and first experimental prototype work support the benefits of the applied blockchain:

- The approach enables producers and agricultural technicians to take part in service design.
- It provides data flow security and can even be included in business models of enterprises.
- It provides utilities for the improvement of internal procedures.
- It allows cost savings by applying smart contracts adapted to the commercial relationships of the network participants.

Because it allows integration with automated installations, implementation expenses are not large.

This paradigm adds fresh demands and problems to be resolved. Software technology needs to be carefully planned in order to be incorporated into businesses without creating additional issues. The user interfaces and action items need to be extremely intuitive and appropriate for the way technicians and farmers operate. The platform must cover the cost of updating and maintaining it. There is a learning curve that needs to be climbed as well. Through the design and creation of a model and experimental platform, which are illustrated through a proof of concept, this study addresses some of these requirements and experimentation demonstrates the model's advantages.

References

[1] Babapoor, A., Hajimohammadi, R., Jokar, S.M. and Paar, M. 2020. Biosensor design for detection of mercury in contaminated soil using rhamnolipid biosurfactant and luminescent bacteria. Journal of Chemistry 2020: 1–8.

[2] Chiang, C.T. and Hsu, C.Y. 2021. A soil yeast count monitor for plant growing applications. IEEE Sensors Journal 21.20: 23510–23517.

[3] Afrad, M.S.I., Monir, M.B., Haque, M.E., Barau, A.A. and Haque, M.M. 2020. Impact of industrial effluent on water, soil and Rice production in Bangladesh: a case of Turag River Bank. Journal of Environmental Health Science and Engineering 18: 825–834.

[4] Boopathy, S., Anand, K.G., Priya, E.D., Sharmila, A. and Pasupathy, S.A. 2021. Iot based hydroponics based natural fertigation system for organic veggies cultivation. 2021 Third International Conference on Intelligent Communication Technologies and Virtual Mobile Networks (ICICV). IEEE.

[5] Ferrández-Pastor, F.J., Mora-Pascual, J. and Díaz-Lajara, D. 2022. Agricultural traceability model based on IoT and Blockchain: Application in industrial hemp production. Journal of Industrial Information Integration 29: 100381.

[6] Shaikh, F.K., Karim, S., Zeadally, S. and Nebhen, J. 2022. Recent trends in internet of things enabled sensor technologies for smart agriculture. IEEE Internet of Things Journal.

[7] Serrano-Calvo, R., Cutler, M.E.J. and Bengough, A.G. 2021. Spectral and growth characteristics of willows and maize in soil contaminated with a layer of crude or refined oil. Remote Sensing 13.17: 3376.

[8] Kumar, R. and Tripathi, R. 2021. Towards design and implementation of security and privacy framework for internet of medical things (iomt) by leveraging blockchain and ipfs technology. The Journal of Supercomputing (2021): 1–40.

[9] Kashyap, B. and Kumar, R. 2021. Sensing methodologies in agriculture for soil moisture and nutrient monitoring. IEEE Access 9: 14095–14121.

[10] Yin, H., Cao, Y., Marelli, B., Zeng, X., Mason, A.J. and Cao, C. 2021. Soil sensors and plant wearables for smart and precision agriculture. Advanced Materials 33.20: 2007764.

[11] Ullo, S.L. and Sinha, G.R. 2020. Advances in smart environment monitoring systems using IoT and sensors. Sensors 20.11: 3113.

[12] Siddique, A., Prabhu, B., Chaskar, A. and Pathak, R. 2019. A review on intelligent agriculture service platform with lora based wireless sensor network. Life 100: 7000.

[13] Caro, M.P., Ali, M.S., Vecchio M. and Giaffreda, R. 2018. Blockchain-based traceability in Agri-Food supply chain management: A practical implementation. pp. 1–4. 2018 IoT Vertical and Topical Summit on Agriculture - Tuscany (IOT Tuscany), Tuscany, Italy, 2018, doi: 10.1109/IOT-TUSCANY.2018.8373021.

[14] Udutalapally, V., Mohanty, S.P., Pallagani, V. and Khandelwal, V. 2020. sCrop: A novel device for sustainable automatic disease prediction, crop selection, and irrigation in Internet-of-Agro-Things for smart agriculture. IEEE Sensors Journal 21.16: 17525–17538.

[15] Merenda, M., Porcaro, C. and Iero, D. 2020. Edge machine learning for ai-enabled iot devices: A review. Sensors 20.9: 2533.

[16] Radoglou-Grammatikis, P., Sarigiannidis, P., Lagkas, T. and Moscholios, I. 2020. A compilation of UAV applications for precision agriculture. Computer Networks 172: 107148.

[17] Chen, M., Mao, S. and Liu, Y. 2014. Big data: A survey. Mobile Networks and Applications 19: 171–209.

[18] Wu, H.T. and Tsai, C.W. 2019. An intelligent agriculture network security system based on private blockchains. Journal of Communications and Networks 21(5): 503–508, oct 2019. doi:10.1109/jcn.2019.000043.

[19] Sharma, P.K., Singh, S., Jeong, Y.S. and Park, J.H. 2017. DistBlockNet: A distributed blockchains-based secure SDN architecture for IoT networks. IEEE Communications Magazine 55(9): 78–85. doi:10.1109/mcom.2017.1700041.

[20] Zhou, L., Wang, L., Sun, Y. and Lv, P. 2018. BeeKeeper: A blockchain-based IoT system with secure storage and homomorphic computation. IEEE Access 6: 43472–43488. doi:10.1109/access.2018.2847632.

[21] Ma, M., Shi, G. and Li, F. 2019. Privacy-oriented blockchain-based distributed key management architecture for hierarchical access control in the IoT scenario. IEEE Access 7: 34045–34059. doi:10.1109/access.2019.2904042.

CHAPTER 9

Software-defined Networking for IoT

Deepa V. and *Sivakumar B.**

‖‖

1. Introduction

With the new conceptual framework of technologies for information and communications, the most popular innovation is Internet of Things (IoT) [1]. The "Internet of Things (IoT)" is a phrase that resulted from the practice of linking embedded items and things to the Internet. The infrastructure of the Internet of Things (IoT) consists of sensing devices, storage, computation, and connectivity, which together form a global and dynamic network architecture. An array of linked smart devices, such as detectors/sensors, smart devices, wearable/portable gadgets, Radio Frequency Identification (RFID) tags and others, are capable of serving as data gathering and broadcasting points with the network. Researchers envision an era in which Internet of Things (IoT) devices will be deployed in great numbers all around us, and create tremendous volumes of data without the direct participation of humans [2]. Nevertheless, in the next years, the field of Internet of Things (IoT) will likely become more appealing to academics as a result of the creation of new application areas that have the potential to significantly enhance our level of life [3].

The IoT encompasses a diverse range of applications, ranging from leisure and athletic pursuits, such as intelligent activity trackers, to mission-critical infrastructure, including healthcare, mobile devices,

Department of Computing Technologies, School of Computing, SRM Institute of Science and Technology, Kattankulathur, Tamilnadu.
Email: dv1018@srmist.edu.in
* Corresponding author: sivakumb2@srmist.edu.in

communication technologies, and smart cities. The development of sensor technologies, cloud networks, and access technology providers, among other factors, has been a significant driving force for the emergence of these applications. As a consequence of this, enormous amounts of data generated by the Internet of Things (IoT) that contain information gathered from real-world sensors have dramatically raised the requirement for storage and computing facilities within IoT ecosystems in order to provide information or services [4]. IoT systems have the potential to manage simultaneous requests when the devices are grouped together in groups of several hundred, thousands, or even millions. This capability is needed by a variety of applications that demand prompt replies [5].

In order to have successful deployments of IoT, diverse communication structures need to be merged together. This involves combining smart gateways so that IoT devices may be connected to the internet. In recent years, the focus of research efforts has shifted towards interlinking the network infrastructure of the Internet of Things with technologies such as edge computing, cloud computing, fog computing, machine learning, and other similar technologies that enhance the possibilities of the IoT [6, 7]. Thus, the administration of IoT networks has become an incredibly challenging problem, and this difficulty will become much more complex in networks beyond 5G, namely 6G, owing to the enormous expansion of connected devices. As a result of these issues, researchers have come up with innovative methods for managing the Internet of Things, such as for security, scalability, energy management [8], load balancing [9], and fault tolerance [10]. In this chapter, the software-defined approach for the IoT is a major technological advancement in the realm of network technologies. The SDN simplifies network management by utilising Open Flow (OF) to physically isolate the network control from the data forwarding components and conceptually centralising it on high-end servers. The SDN framework proposes a hierarchical structure that comprises three distinct planes, namely the application plane which encompasses Application Programming Interfaces (APIs), the control plane which involves a controller, and the data forwarding plane. The SDN controller bears the responsibility of upholding the complete processing network state and programmable APIs enable the control plane to prioritise, classify, and deprioritise network traffic in a systematic manner.

The goal of software-defined networking (SDN) is to make the architecture of the network more flexible with intelligence so that it can dynamically adjust to changes in the network environment at run time [11, 12]. Because of the high level of changeability inherent in an Internet of Things network, such as limits on battery life, computing power, and storage capacity, the network is required to adapt to specific demands as shown in Fig. 1. In light of the fact that the SDN architecture makes the administration of IoT networks a great

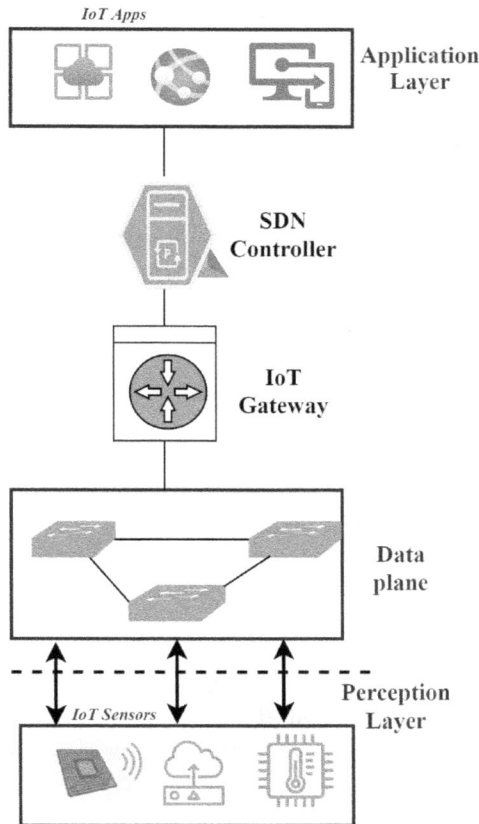

Figure 1. Architecture of SDN-based IoT.

deal simpler, a significant amount of research effort is being directed in this particular area. There have been a number of research conducted to study the various IoT reference architecture models that rely on SDN for both current and future IoT deployments.

The combination of IoT with SDN (SDN-IoT) makes decoupling the control plane from the data plane and allows devices to be connected via the Internet. SDN offers the ability to make the IoT network scalable and programmable while also providing possible answers for the highlighted problems associated with IoT management. This can be accomplished without disrupting the fundamental architecture of any current implementations. In this chapter, the section furnishes an overview of Resource Management and Allocation, SDN Enabled IoT Scenario, and securing the Communication using Blockchain Technology.

2. Background

Characteristics in IoT

With developments in technology (such as internet services, cloud computing, and terms of implementing, amongst others) and the incorporation of the Internet of Things (IoT) into a wide diversification of fields, a great number of definitions describing the IoT concept have been published by a variety of organisations and researchers [13]. Despite the fact that there are many different ways in which it can be defined, there are a few characteristics that are shared by all Internet of Things (IoT) systems.

The domain of research and development has witnessed a growing emphasis on the Internet of Things (IoT) in recent times. The Internet of Things (IoT) refers to a system of interconnected physical objects, devices, and equipment that are capable of communicating with each other through the internet. The establishment of this connection facilitates the acquisition and exchange of information in a synchronous manner. The topic of ensuring security for IoT has become increasingly urgent due to its exponential growth. There are several obstacles to overcome in order to ensure the safety of the Internet of Things (IoT), including the difference between connected gadgets and communication protocols, the high number of possible entry points, and the constrained computing power and memory of many IoT devices [14]. The security of IoT systems must be addressed at every layer, beginning with the gadgets themselves and moving all the way up to the applications that make use of the data that the sensors gather. The Internet of Things is composed of embedded technologies, each of which has limited access to certain resources. The presence of variability in sensors, connections, software, operations, and/or information is the primary trait that distinguishes it from other types of cloud environments [15]. In addition to this, the Internet of Things system's network connection is dynamic and made up of a vast number of devices, some of which are permanently linked while others connect and detach intermittently over the course of time. This is due to efforts to save energy or fluctuating circumstances on the network.

The difficulties with flexibility and interoperability might be heterogeneity in an Internet of Things. Because of this, it is hard to administer the network and exercise control over the vast and varied collection of devices that are connected to it. These devices use a variety of communication protocols to share different kinds of information with one another. In addition, the IoT system must be capable of adding new client devices, services, and functionalities without degrading the quality of the already existing network services.

SDN-IoT (Software-defined Networking - Internet of Things) Approaches

IoT technology is currently experiencing significant changes due to the exponential increase in the quantity of various kinds of devices that possess the capability to connect the internet.

Protecting such complicated networks and the many ways in which they may be accessed is a significant challenge that might result in an increased risk [16]. The SDN paradigm offers an alluring option for controlling the Internet of Things services that have lately come into focus [16, 17]. It is able to keep track of traffic in an intelligent manner and make use of network resources that are not commonly used [18]. Separating the control plane, which makes decisions about how to handle traffic, from the data plane is one of the benefits of software-defined networking (SDN) (real traffic transmission processes to desire locations). The decoupling encourages the abstraction of lower-level network operations into higher-level facilities, which makes it much simpler to handle the chores associated with network management. The planned outcome of development is a noteworthy enhancement of the network's capacity, thereby rendering it more amenable for systems to fortify themselves against a potential data breach involving the Internet of Things (IoT). It would minimise inefficiencies, allowing for efficient processing of information provided by IoT without placing significant stress on the network, most notably the Wi-Fi network [19].

The combination of IoT with SDN (SDN-IoT) [20] creates the process of obtaining information, processing information, making choices, and decisions into action. The implementation of SDN-IoT enables monitoring and the management of network assets, as well as access management that varies according to user, organisation, device, and implementation. As a result of the development of SDN-IoT, networks stand to profit from an increase in their power to govern other networks and focusing on demand may become more scalable with the help of SDN.

Resource Management and Allocation

Recent challenges associated with resource management and allocation in IoT systems, as well as new research into potential solutions to the issues, shows us that in IoT systems, almost every resource, including multiple servers, transmission systems, and wireless resources, must be shared since they are utilised by a large number of activities and requests. It is of the utmost importance that the process of sharing designs is both efficient and successful. The computational time complexity, high propagation delay, narrowing of coverage, allocation of resource constraints, low power efficiency, security issues, and Quality of Service (QoS) requirements that can be caused by inefficient resource allocation are the most obvious issues that arise on resource management.

The Problems with resource management and allocation are crucial for resource-constrained systems like the conventional Internet of Things, the analytical Internet of Things, and machine-to-machine interaction (M2M). Researchers typically use artificial intelligence (AI), Virtual machine (VM), machine learning (ML), and optimisation approaches in recent studies for resource allocation in these systems. These approaches aim to create high-efficiency, low-latency frameworks that can serve a large number of users and enable full use of resources available. The difficulties that may be encountered while using IoT systems are addressed in a variety of research and efficient strategies for overcoming these difficulties are established. Here, the Offloading operations make it possible for resource-constrained Internet of Things devices to carry out a variety of sophisticated and time-consuming tasks. IoT devices have a limited battery life, thus in order to run complex application functions, they rely on cloud computing or edge computing. The difficulty of picking one edge server to be offloaded among numerous edge servers may be solved by allocating resources. The utilisation of fog computing or edge computing may both be used for computational offloading. The ability to do more sophisticated tasks is one of the benefits offered by cloud computing, among other advantages.

The fact that it is centralised and removed from the IoT devices makes it susceptible to breaches in security and data quality, as well as possible delays in processing times. As a result, processing in real-time is necessary, and the structure must be dispersed in order to accommodate applications and reliable data storage.

SDN-IoT Based Resource Allocation

It has also come to light that the SDN controller is already familiar with the current condition of the edge server's resource use. The implementation of a regulation that dynamically updates the flow table in the SDN controller. This rule determines how resources will be distributed from a particular edge server based on the application needs of each class. The OpenFlow switch's flow table is updated to include the newly produced rule from when it has been created. The topology effectively partitions the control area from the data segments, thereby enabling adaptable network management and enhancing the performance of time-sensitive applications. The Identification of resources and allocation of those resources are both critical steps for computing in the IoT. At the step known as resource verification, the SDN controller collects information on the edge server's available resources. In order to achieve the effective execution of each application in resource allocation, it is necessary to have a solid understanding of the capabilities involved in resource allocation, resource information, and resource consumption.

Fog-IoT Structure with SDN

For the purpose of supporting IoT networks and applications, the system that is being suggested integrates the SDN paradigm with the fog computing notion.

As illustrated in Fig. 2, the system relies on three tiers of hardware devices and network components. The initial layer of devices, representing either the Internet of Things (IoT) devices or the sensor layer, consistently generates data that necessitates transmission throughout the network. The second layer is commonly referred to as the fog layer, which assumes the role of deploying fog nodes. The aforementioned nodes facilitate a channel for data offloading and offer the diverse benefits of fog computing for the Internet of Things (IoT) network. The remote cloud unit represents the uppermost layer of the cloud. The cloud infrastructure for the Internet of Things offers assistance for a range of services and protocols related to IoT. The integration and connection of the Internet of Things cloud with other networks can be assisted by the service provider.

System Structure

The system structure is comprised of a number of stages; the authentication of the IoT node is the first stage since it is required to be approved. In order to get authorisation, the IoT node interacts directly with the IoT cloud to carry out the authentication procedure and name the device that has to be approved.

The identification of the address is the next procedure. The cloud makes a call to the service provider in order to find out where the IoT is located. To achieve this objective, the service provider contacts the SDN Orchestrator, who invests money to identify the IoT. In addition to this, the SDN orchestrator makes an estimate of the routing table, which includes a variety of alternative routing pathways between the IoT node and cloud, and it finds all Open Flow (OF) switches that are devoted to this connection.

The system relies heavily on the idea of resource utilisation, and as a result, it makes full use of all of the resources that are at its disposal. As a consequence of this, the SDN controller enables selected OF switches to do part of the processing and computing functions for the IoT data that has been transmitted after the fog level. The estimation of switches by the SDN controller is predicated upon the evaluation of specific factors that inform the availability of resources. The criteria under consideration includes IoT traffic, transit traffic, traffic access type, time delay limits, processing capability for serving IoT data, and the current condition of OF switches in terms of traffic and resources.

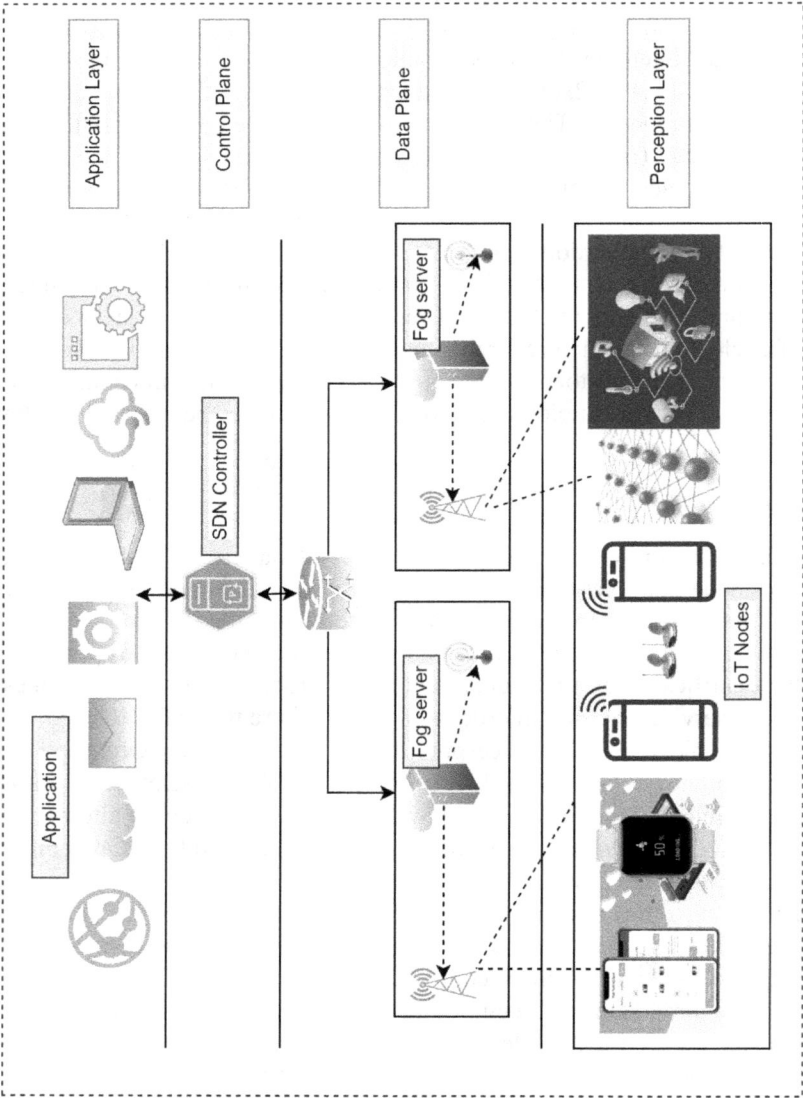

Figure 2. Fog-IoT Structure with SDN.

The feasibility of activating the IoT data segment resources of OpenFlow (OF) switches is ascertained by the software-defined networking (SDN) controller, which optimises the preceding parameters and subsequently communicates the selection to the switches. The orchestration process involves the creation of a virtual machine on selected OpenFlow switches, which will serve as a platform for data processing. The subsequent stage involves the migration of the database, which will occur upon the transfer of the foundation for facilitating the IoT group to the OF switch by means of the service provider of the IoT Cloud. The network's functionality remains uninterrupted, and OpenFlow switches persist in gathering and synchronising data emanating from Internet of Things (IoT) devices with cloud-based storage.

Intelligent Resource Allocation

There have been a few studies conducted of integrating the technologies of SDN, Fog computing, and IoT. As a result, with unique perspectives on this problem which has been investigated by taking into account many things including architecture, virtualisation, requirements, and standards, to acquire a new accomplishment in the integration of SDN, IoT, and fog computing technique.

In [20], the authors focused on the infrastructure setup and execution of a 5G open networking solution, the methodology of SDN/NFV-based 5G and IoT, as well as how AI/ML interacts with and learns from 5G/IoT. This article reviewed a number of sectors that have been improved by using AI/ML/DL methodologies to function as a resource or in order to provide to other sectors.

In [21], a framework for the construction of the Internet of Things was presented. This framework utilises the edge computation layer in the fog nodes. The SDN is implemented inside the framework, complete with a centralised controller and OpenFlow distributed switches. The switches exhibit a restricted capability in terms of both storage and memory. The operation of the network is based on an algorithm for unloading data, also known as load distribution. This method delegated all processing and computational activities to specific OpenFlow switches, which make use of idle resources. The technique that has been outlined provides a number of benefits for the IoT, including the reduction of delay and the increase in resource efficiency. The suggested technique has been shown to attain a high level of efficiency in terms of delay and resource usage, as shown by the experimental results obtained in the testing environment during the evaluation of the method.

In [22], a two-level hierarchical control system, consisting of a master and a slave, was established to solve the problem of bottlenecks that can be caused by the current level of the single-control plane, which is located on the remote

IoT gateways. This is due to the fact that the current level of the single-control plane is on the remote IoT gateways. To achieve optimal control performance, a slave controller placement strategy has also been proposed as an option. In the end, a number of different tests were carried out in order to evaluate how effective the suggested MATLAB approach was. The results of comparative analyses indicate that the proposed approach reduces the crucial latency of managing the Internet of Things (IoT) by approximately 30.56%.

In [23], an innovative software-defined application was proposed. This application makes it possible to build an application which facilitates the construction of a malleable and responsive framework for exchanging information within the context of 5G-IoT. The infrastructure is designed using the SoftAir 5G system and is intended to improve the efficacy of data transmissions. In order to accomplish this goal, SoftAir partitions a software-defined wireless architecture into a control plane and a data plane. This allows for a more efficient coordination between millimetre wavelength (mmWave) and remote radio heads (RRHs), which are required for Internet of Things access. Subsequently, a novel architectural framework comprising SD-GWs and SoftAir local IoT controllers was released in order to facilitate the efficient administration of diverse IoT applications and the heterogeneity of IoT systems.

In [24], the goal of fog computing is to supplement time-sensitive Internet of Things (IoT) applications that need to meet low-latency requirements by supplying resources to cloud data centres that are located at the network's edge. In order for the IoT-Fog network to successfully carry out the requests made by its users, it is essential to shield its resources and scheduling services from any potential threats that may arise. The implementation of suitable scheduling algorithms is crucial for effectively addressing the requirements of user applications and maximising the utilisation of IoT-Fog resources. SDN switches and controllers are able to function as cloud gateways and fog gateways in IoT-Fog networks. On the other side, SDN switches and controllers are more prone to being attacked by a wide range of threats, which makes the SDN controller a bottleneck and makes it simple to regulate plane saturation. This approach provides safeguarding measures for scheduling services to mitigate potential threats such as distributed denial of service (DDoS) and port scanning. The S-FoS technique is utilised for the purpose of enhancing security and optimising performance. It involves the implementation of anomaly detection algorithms based on fuzzy logic to accurately identify the origin of attacks and subsequently prevent malicious requestors from accessing the system. The experimental findings indicate that S-FoS exhibits a potential enhancement of 31% and 18% in response time, and 9% and 4% in network utilisation, in comparison to the NSGA-II and MOPSO algorithms, respectively. The aforementioned enhancements could be attained through the manipulation of attack rates, the type of IoT devices, and fog devices.

SDN Enabled IoT Scenario

Figure 3 depicts the framework and methodology of the SDN-enabled IoT Scenario in fog networks. The Internet of Things (IoT) may employ a dynamic structure to enhance network security by leveraging user contextual information, thereby mitigating and addressing cybersecurity threats at the network edge. The security framework is furnished with response and monitoring tools, and it employs novel algorithms and approaches to analyse and correlate threats from diverse sources, thereby enhancing security measures.

- *Users/Fog Nodes*: The Users/Fog Nodes offer the more powerful compute and processing resources for IoT devices are provided by the node. In contrast, the fog node facilitates the proximity of the cloud to the object. Nodes in the fog and at the network's edge may take the form of switches, gateways, computers, or I/O devices. These nodes are responsible for providing storage, computation, and connection services. In addition, switches and controllers for SDN networks may be included inside them.

- *Encrypted Flow rules*: If unauthorised individuals gain entry, the information retrieved from diverse Internet of Things gadgets may comprise confidential data and potentially result in consequential ramifications. To clarify, the architecture has the capability to alter the data and does not ensure data security for devices that possess identical source files and hash keys. Prior to organising the data into blocks, the system employs encryption techniques such as AES-256, AES-128, or RSA to safeguard the confidentiality of the data and to fortify the system's protective capabilities. The encryption technique employed can be modified based on the protocols governing key exchange and device configurations. Stated differently, the efficacy of encryption may be restricted by unauthorised access.

- *SDN Controller/IoT Controls*: The functionality of these modules is attributed to the utilisation of resources and security enablers. Both the Internet of Things (IoT) networks and software-based networks are accompanied by run-time operations that are implemented in their respective systems. To effectively handle connectivity across the physical infrastructure and the underlying virtual layer, a collection of dispersed SDN controllers that can interact with network components based on SDN is necessary. A fog node is capable of executing the tasks of an

SDN controller, in addition to performing network operations and implementing flow rules for a diverse range of fog nodes and users.

- *SDN/NFV Management/Orchestrator*: The virtualisation of Internet of Things access network devices is a task assigned to the network layer. The concept of software-defined networking (SDN) enables the administration, monitoring, and management of networks to be automated and dynamic. This is achieved by separating the control operations from the data forwarding process. This layer is comprised of the SDN controllers as well as the SDN switches (i.e., OVS (Open Virtual Switches), OFS (Open Flows Switches)). The software-defined network controller (SDN controller) has the flexibility to be centralised or distributed depending on the needs of the network. The Security Orchestrator is responsible for making the decision of which enablers should be successfully implemented, taking into account the prerequisites

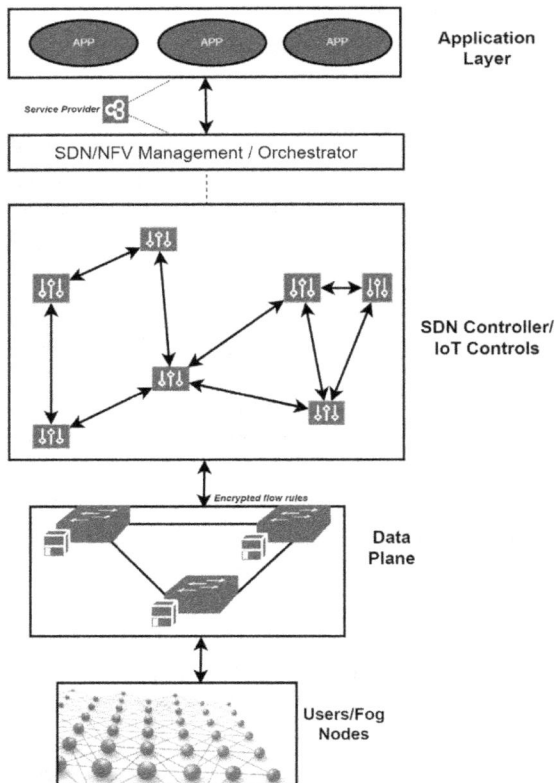

Figure 3. Framework of SDN Enabled IoT scenario.

for security, the resources that are already available in the infrastructure, and the optimisation criteria.

- *Service Provider*: The Security Provider/Enabler with Interpreter is utilised in the context of Software-Defined Networking (SDN) and Network Function Virtualisation (NFV) to identify the IoT-specific enablers and provide the implementation of necessary capabilities.

- *Application Layer*: The component is the architectural element that generates the primary attention from the end user. The application layer, situated at the top of the protocol stack, is accountable for managing and manipulating the data and information obtained. The application layer depicts the various applications or services that are part of the Internet of Things, including but not limited to Industrial IoT, Smart Households, Smart Buildings, and Smart Healthcare.

3. Security and Privacy of SDN-IoT

With the unique characteristics of SDN-IoT, it is possible to directly use any of the current technologies for ensuring secure communications. This section examines the diverse range of IoT security solutions and evaluates the efficacy in safeguarding a computing environment.

Topology Management

For the functioning of IoT nodes, one of the operations that is considered to be amongst the most vital and crucial is known as Topology management. The creation of a graph that illustrates an abstract picture of the network nodes and the communication linkages between them is the objective of the topology control process. The primary use for this abstract representation is in making routing choices with the objectives of minimising congestion, cutting down on load, minimising latency, and optimising energy consumption.

Flow Management

Even though focus primarily on flows from devices to servers, due to the fact that this is the most typical scenario for SDN-based communications. The potential IoT flow within software-defined IoT pertains to the transfer of data between IoT devices placed in distinct partitions. The IoT flow that could potentially exist in software-defined IoT is the data flow between IoT devices that are located in different partitions. Here, Inter-region communication is required in order to plan out this particular kind of Internet of Things flow. The controllers exhibit collaborative functionality by utilising the corresponding switches to redirect the flow initiated from one area to an alternative entry point situated in a distinct area. In contrast to the communication and controller resources available in the region partition's path from the controller to receiver, the path from controller to receiver within the partition exhibits lower throughput and delay. The reason for this phenomenon is attributed to the high density of Internet of Things (IoT) devices within the partition, which leads to an increased requirement for communication resources and a more intensive processing load on the associated controller.

4. Security Attacks in Communication

Attacks Based on Node Availability

A rogue IoT node might appear as a legal one in order to share and gather the data that is created by other IoT devices for the aim of committing further acts of harmful activity. The rogue Internet of Things nodes have the ability to abuse the data of users or to give harmful data to surrounding nodes, which may disturb the behaviors of those nodes. The complexity of trust management across the many schemes in the IoT might make it difficult to find a solution to this challenge. On the other hand, a trust measurement-based strategy might be used to identify rogue nodes in IoT settings, which could provide some protection against potential security threats.

Attacks Based on Confidentiality

The IoT scenario incorporates a broad range of gadgets and sensor systems that are associated with a number of different SDN controllers. There is no reliable system in place that can determine when and how to place confidence in IoT devices. Given the absence of a trustworthiness evaluation, individuals utilising Internet of Things (IoT) services must evaluate whether it would be financially advantageous to abstain from certain IoT services. Establishing a trustworthy environment among the Internet of Things (IoT) devices is a crucial element in ensuring secure settings and preserving the integrity of IoT services. As a result, creating an environment of trust among Internet of Things devices is one of the most important aspects of setting up safe settings in order to maintain the integrity of Internet of Things services. There have been various situations, such as social media platforms on the internet, in which trust models that are based on reputation have been effectively implemented. To create a trust model in the IoT that is based on reputation, they have to address how to maintain availability, and reliability, and how to avoid adverse malfunctions, how to correctly identify malicious behaviour, and how to automate assembling a security framework based on reputation in large-scale networks.

Attacks Based on Authenticity

Authenticity is a fundamental need for ensuring the safety of Internet of Things devices. However, many of the devices that make up the IoT do not have the space or processing capacity to carry out the encryption processes that are necessary for security mechanisms. These devices with limited resources have the ability to outsource their complex calculations and storage needs to a fog device, which will then carry out the authentication procedure. The strategy ensures that all communications are kept private by using a public key together with multicast authentication.

Attacks Based on Integrity

The messages created by IoT devices are sent to the fog nodes that are logically the closest to them. With the Internet of Things devices, it poses a challenge to handle a vast amount of data. The data is first segmented into different pieces before being sent to a number of fog nodes for processing. At this stage, the contents of the data need to be evaluated without the data itself being exposed. It is essential that the data's integrity be maintained whenever it is combined with previously transmitted and processed information. Encryption methods or masking techniques are able to encrypt or decrypt data on an Internet of Things device with devices that have limited resources.

Attacks Based on Non-Repudiation

Data aggregation, Key management, and verified computing are a few of the additional security difficulties that must be overcome. Despite this, the distinct qualities of fog computing have the potential to make a contribution towards resolving concerns relating to the user's privacy and data security in IoT contexts. The Internet of Things (IoT) systems are susceptible to a variety of attacks, including Denial of Service (DoS) attacks and malware-based attacks, which are among the most frequently encountered types of attacks. Additionally, fog computing may serve as a constituent of the security mechanism that guarantees the immunity of IoT services against such attacks. In the event of a Denial of Service (DoS) attack, the extensive distribution of fog nodes may potentially aid in preserving the resilience of IoT services.

5. Secure Communication using Blockchain Technology

The combination of fog computing and the IoT makes it possible to provide services at the network's outermost layer, such as computation and storage, which reduces response latency and enhances the user experience. The dynamic properties in terms of time and location of the data that is detected by IoT devices define the deployment of fog nodes as a programming problem with an uncertain solution. The purpose of the optimisation effort is to reduce the amount of power that is sent, and blockchain technology [25] is given as a means of finding the best answer. In order to assess the algorithm, we execute a series of simulations, and the results demonstrate that it offers superior performance compared to the baseline approaches.

The IoT's growth has accelerated quickly in recent years, and security concerns have grown as a result. The insecure nature of the communication that takes place between different Internet of Things devices is one of the security issues that might develop on the nodes or communication channels/pathways.

6. Conclusion

In this chapter, the design and execution of an IoT system have been carried out with the assistance of SDN technology so that the findings may be compared. The various communication methods and intelligent resource allocation algorithms for the IoT system are discussed here. In the meanwhile, the Internet of Things system that incorporates blockchain technology makes use of a decentralised blockchain network's platform as a trusted central authority. In addition, the data that users want to store and retrieve from the blockchain network may be done with the use of intelligent contracts. It is possible that the IoT system does make use of blockchain technology, which has a higher level of security.

References

[1] Babun, L., Denney, K., Celik, Z.B., McDaniel, P. and Uluagac, A.S. 2021. A survey on IoT platforms: Communication, security, and privacy perspectives. Computer Networks, Volume 192.

[2] Harjula, E. et al. 2019. Decentralized Iot edge nanoservice architecture for future gadget-free computing. IEEE Access 7: 119856–119872.

[3] Qiu, T., Chen, N., Li, K., Atiquzzaman, M. and Zhao, W. 2018. How can heterogeneous Internet of things build our future: A survey. IEEE Communications Surveys & Tutorials 20(3): 2011–2027.

[4] Marjani, M., Nasaruddin, F., Gani, A., Karim, A., Hashem, I.A.T., Siddiqa, A. and Yaqoob, I. 2017. Big IoT data analytics: architecture, opportunities, and open research challenges. IEEE Access 5: 5247–5261.

[5] Stankovic, J.A. 2014. Research Directions for the Internet of Things. IEEE Internet of Things Journal 1(1): 3–9.

[6] Yugha, R. and Chithra, S. 2020. A survey on technologies and security protocols: Reference for future generation IoT. Journal of Network and Computer Applications, Volume 169.

[7] Gubbi, J., Buyya, R., Marusic, S. and Palaniswami, M. 2013. Internet of things (IoT): A vision, architectural elements, and future directions. Future Generation Computer Systems 29(7): 1645–1660.

[8] Li, J., Herdem, M.S., Nathwani, J. and Wen, J.Z. 2023. Methods and applications for Artificial Intelligence, Big Data, Internet of Things, and Blockchain in smart energy management. Energy and AI, Volume 11.

[9] Verma, R. and Chandra, S. 2023. HBI-LB: A Dependable Fault-Tolerant Load Balancing Approach for Fog based Internet-of-Things Environment. J. Supercomput. 79: 3731–3749.

[10] Kiadehi, K.B., Rahmani, A.M. and Molahosseini, A.S. 2021. Increasing fault tolerance of data plane on the internet of things using the software-defined networks. PeerJ Computer Science 7: e543.

[11] Alabbad, M. and Khedri, R. 2021. Configuration and governance of dynamic secure SDN. Procedia Computer Science 184: 131–139.

[12] Sufiev, H., Haddad, Y., Barenboim, L. and Soler, J. 2019. Dynamic SDN controller load balancing. Future Internet 11(3): 75.

[13] Lin, J., Yu, W., Zhang, N., Yang, X., Zhang, H. and Zhao, W. 2017. A survey on internet of things: architecture, enabling technologies, security and privacy, and applications. IEEE Internet of Things Journal 4(5): 1125–1142.

[14] Yang, X., Li, Z., Geng, Z. and Zhang, H. 2012. A Multi-layer Security Model for Internet of Things. Internet of Things, Berlin Heidelberg: Springer, pp. 388–393.

[15] Masoudi, R. and Ghaffari, A. 2016. Software defined networks: A survey. Journal of Network and Computer Applications 67: 1–25.

[16] Kalkan, K. and Zeadally, S. 2017. Securing internet of things (IoT) with software defined networking (SDN). IEEE Communications Magazine.

[17] Sultana, N., Chilamkurti, N., Peng, W. and Alhadad, R. 2018. Survey on SDN based network intrusion detection system using machine learning approaches. Peer-to-Peer Networking and Applications, pp. 1–9.

[18] Turner, S.W., Karakus, M., Guler, E. and Uludag, S. 2023. A promising integration of SDN and Blockchain for IoT Networks: A survey. IEEE Access 11: 29800–29822.

[19] Chin, W.L., Ko, H.A., Chen, N.W., Chen, P.W. and Jiang, T. 2023. Securing NFV/SDN IoT using Vnfs over a compute-intensive hardware resource in NFVI. In IEEE Network, doi: 10.1109/MNET.135.2200558.

[20] Lin, B.S.P. 2021. Toward an AI-enabled SDN-based 5G & IoT network. Netw. Commun. Technol. 5(2): 1–7.

[21] Khakimov, A., Ateya, A.A., Muthanna, A., Gudkova, I., Markova, E. and Koucheryavy, A. 2018. IoT-fog based system structure with SDN enabled. In Proceedings of the 2nd International Conference on Future Networks and Distributed Systems (ICFNDS '18). Association for Computing Machinery, New York, NY, USA, Article 62: 1–6.

[22] Ren, W., Sun, Y., Luo, H. and Guizani, M. 2018. A novel control plane optimization strategy for important nodes in SDN-IoT networks. IEEE Internet of Things J. 6: 1–14.

[23] Oquendo, L.T., Lin, S.C., Akyildiz, I.F. and Pla, V. 2019. Software-defined architecture for QoS-aware IoT deployments in 5G systems. Ad Hoc Netw. 93: 1–11.

[24] Javanmardi, S., Shojafar, M., Mohammadi, R., Persico, V. and Pescapè, A. 2023. S-FoS: A secure workflow scheduling approach for performance optimization in SDN-based IoT-Fog networks. Journal of Information Security and Applications, Volume 72.

[25] Fakhri, D. and Mutijarsa, K. 2018. Secure IoT communication using Blockchain technology. pp. 1–6. 2018 International Symposium on Electronics and Smart Devices (ISESD), Bandung, Indonesia.

CHAPTER 10

Dynamic Threshold-based DDoS Detection and Prevention for Network Function Virtualization (NFV) in Digital Twin Environment

Sriramulu Bojjagani*, N. Surya Nagi Reddy,
Siva Sathvik Medasani, Mohammad Umar, Ch. Avinash Reddy
and *Neeraj Kumar Sharma*

1. Introduction

With the increasing reliance on cloud-based services and the widespread adoption of Network Function Virtualisation (NFV), the threat landscape for Distributed Denial of Service (DDoS) attacks has become more complex and challenging. DDoS attacks can cause significant disruptions, leading to service unavailability, financial losses, and reputational damage for organisations [1–18]. Therefore, developing effective mechanisms for early detection and the prevention of DDoS attacks in the context of NFV is crucial. NFV provides a flexible and scalable infrastructure for deploying and managing network services by virtualising network functions. However, the virtualised nature of NFV introduces new vulnerabilities and attack vectors that DDoS attackers can exploit. Traditional DDoS detection and prevention techniques designed for physical networks may not be directly applicable in the NFV environment.

Cyber Security Lab, Department of Computer Science and Engineering, School of Engineering and Sciences (SEAS), SRM University-AP, Amaravati, Andhra Pradesh, 522240, India.
* Corresponding author: sriramulubojjagani@gmail.com

1.1 Network Function Virtualisation (NFV)

NFV is a concept of architectural framework; the main aim of NFV is to transform traditional networking hardware components into software-based applications. In conventional networks, network functions such as firewalls, routers, load balancers etc., are implemented on specialised hardware devices. This approach often leads to complex and inflexible networks, making introducing new services or scaling existing ones difficult. NFV addresses these challenges by leveraging the principles of virtualisation to create virtualised network functions (VNFs) [4].

1.1.1 Architecture of NFV

Only two levels of the architecture of NFV is mainly divided into four layers:

Network Function Virtualisation Infrastructure (NFVI): This layer represents the underlying hardware resources, including servers, storage devices, and networking infrastructure, which form the foundation for hosting, managing and executing virtualised network functions (VNFs). NFVI includes hardware resources, a virtualisation layer or hypervisor, and virtual resources. Through the virtualisation (hypervisor) layer, hardware resources such as computation, storage, and networking provide processing, storage, and communication to VNFs. Computing, storage, and networking are abstracted from the hardware layer by the virtualisation layer and made available as virtual resources. Virtualised Network Function (VNF): These are represented as software instances running on the NFVI. VNFs can include network functions such as firewalls, routers, load balancers, or intrusion detection systems. For example, it can also be called DHCP server VNF when the DHCP server has been virtualised and Firewall VNF. When the router has been virtualised, it can be referred to as Router VNF; when a base station is virtualised, it is referred to as Base Station VNF [2].

Operation Support System/Business support system (OSS/BSS): OSS covers all aspects of network administration, fault management, configuration management, and service management. Among other things, BSS manages orders, products, and clients. Using standard interfaces, an operator's decoupled BSS/OSS can be integrated with the NFV Management and Orchestration in the NFV architecture. The Management and Orchestration (MANO) Layer contains two components: Virtualised Infrastructure Manager (VIM) component controls which manage the virtualised resources within the NFVI. It communicates with the NFV-I and provides interfaces for VNF managers to manage the virtual machines, virtual networks, and storage resources required for VNF deployments. VNF Manager (VNFM) is responsible for the lifecycle management of VNFs. It interacts with the NFVO and VIM to deploy, monitor, and scale VNF instances according to service requirements. It acts as the central orchestrator of the NFV architecture. The NFVO receives service requests and translates them into specific VNF deployments and configurations across the NFVI. It coordinates with the VIM, NFV-I, and others.

1.2 Distributed Denial-of-Service (DDoS) Attack

Distributed Denial-of-service (DDoS) is a cyber threat that mainly affects system availability. Multiple systems of attackers are used to flood a large amount of traffic to a targeted website or network, overwhelming it and causing it to be unusable to its users. In DDoS attacks, the attacker typically controls a network of compromised computers called a botnet. These compromised computers, often infected with malware, are used to send massive traffic to the target simultaneously. This flood of traffic consumes the target's resources, such as bandwidth, processing power, or memory, causing it to slow down or sometimes crash.

DDoS botnets play an essential role in DDoS attacks. A botnet is a collection of hundreds or thousands of machines, often known as zombies or bots, that hackers have commandeered. The attackers will harvest these computers by identifying weak places they may infect with malware via phishing attacks, malware attacks, and other techniques for mass infection. The hijacked machines may be ordinary home or business PCs, DDoS devices, or even an army of hacked CCTV cameras (the Mirai botnet is notorious for constructing such an army), and their owners are probably oblivious of the situation because they continue to function normally in most other respects. DDoS assaults fall into one of three categories, which are primarily defined by the sort of traffic they direct at their targeted systems:

1. *Volume-based attacks*: utilise a lot of fake traffic to overburden a server or website's resources. Examples include TCP/UDP floods and ICMP floods. The size of this traffic is measured in bits per second (bps).

2. *Protocol or network-layer DDoS attacks*: Sends many packets to your chosen infrastructure management tools and network infrastructures. PPS (packets per second) units measure the scale of these protocol attacks, including SYN floods and Smurf DDoS.

3. *Application-layer attacks*: These attacks target specific applications or services on the target server. They aim to exhaust server resources by exploiting vulnerabilities in the application layer. Examples include HTTP floods and DNS amplification attacks. The units used for measuring are RPS (request per second).

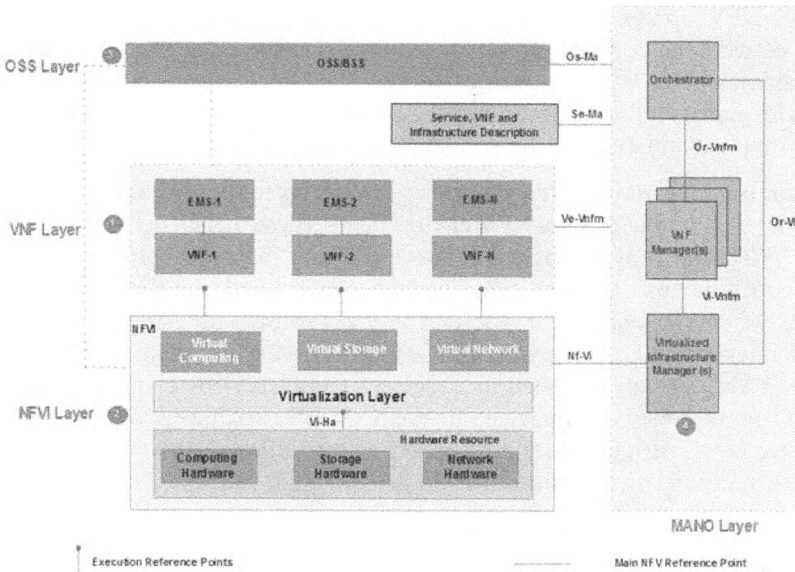

Figure 1. Architecture of Network Function Virtualisation (NFV) [15].

Some of the essential methods applied in all DDoS attack types include:

1. *Spoofing* [1]: The act of an attacker altering or concealing header data that should identify the source of an IP packet is known as "spoofing." Because it cannot see the packet's trustworthy source, the victim cannot prevent attacks from coming from it.

2. *Reflection*: The attacker may use a false IP address to make it look like a packet came from the intended target while sending it to a third-party system that responds to the victim. As a result, it is far more challenging for the victim to identify the attacker in an attack.

3. *Amplification*: It is possible to mislead some internet services into responding to packets with several or massive packets [4].

1.3 System Architecture

The system architecture involves several key components that work together to enable the virtualisation and management of network services. Here's an overview of the entire system architecture. Infrastructure includes hardware resources such as servers, storage devices and network equipment; these resources provide the foundation for the whole system that runs virtually. The system then has a Hypervisor in which the clusters and the VMS are placed; these all are located in the virtualisation layer. In the management layer, the

VIM, NFV orchestrator, and VNF Manager will be present, followed by the network layer Physical Network Infrastructure: The physical network infrastructure includes routers, switches, and other network devices that provide connectivity and transport for network traffic between the virtualised environment and external networks.

Virtualised Network Infrastructure: This refers to the virtual networks and overlays created within the NFV environment. These virtual networks connect the VNFs and provide logical connectivity and isolation between different network functions and services. The topmost layer will be the service layer. Network services are the end-to-end services offered to users or customers, such as virtual private networks (VPNs), software-defined wide-area networks (SD-WAN), or network security services. These services comprise multiple VNFs orchestrated and managed by the NFV infrastructure. On the top of the service layer, various nodes are created where traffic is stored and analysed. These nodes play a significant role in detecting the attacks.

Our key contributions to this chapter are the following:

1. NFV is an architectural framework and concept that aims to virtualise and consolidate various networking tasks and functions that traditionally rely on dedicated hardware devices.

2. The papers have proposed a methodology for state-of-the-art research to mitigate DDoS in the NFV environment. However, in this chapter, we are the first to attempt to integrate NFV and OpenStack (An open-source software platform providing a cloud computing infrastructure) to detect and prevent DDoS attacks in the Digital Twin Environment.

3. Research ideas until now focussed on employing multiple routers or creating a virtual server that acts as a filter to control traffic. However, this chapter addresses the attempts to control the traffic with only some additional variables to optimise resource utilisation and minimise the cost. We were the first to use a dynamic threshold-based algorithm to control network traffic.

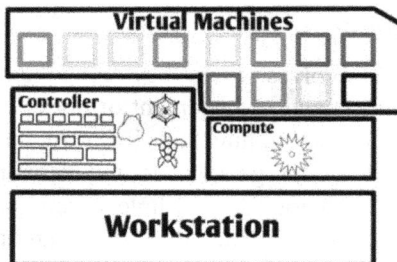

Figure 2. System architecture.

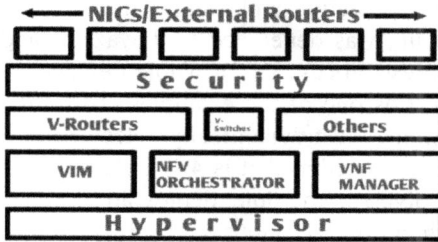

Figure 3. OpenStack architecture within controller.

2. Related Work

Many studies have focused on preventing DDoS attacks in environments like traditional networks, SDN, and NFV. Based on the type of Networks, the researchers have developed different kinds of models in multiple settings for detecting and preventing DDoS attacks. In light of this, researchers have become increasingly interested in virtual networks and virtual environments as part of this tremendous change from traditional networks to virtualisation. Girma et al. [4] provided a hybrid mitigation technique for virtual networks to identify DDoS assaults in cloud computing settings at the network and host levels. They used entropy and covariance matrices. Li, Wang [5] Defence architecture against DDoS assaults is called CODE, and it uses network virtualisation and software-defined networking to implement its functionalities to identify and stop attacks. In particular, the framework can ease the strain on nodes whose traffic exceeds the maximum their self-defence mechanisms allow. The framework also includes a technique for managing resources online that balances participants' fairness and resource consumption efficiency across the board. Of course, this framework from the present still has certain underdeveloped areas. For example, it is designed for SYN flood attacks and does not consider broader DDoS attacks. Our proposed system can mitigate almost all types of traffic-related attacks, which deal mainly with the availability of the servers.

Yang [10] proposed a methodology called VFence that mitigates the DDoS attacks in the NFV. Using Network Function Virtualisation (NFV) technology, VFence protects against DDoS attacks by intercepting packets, confirming their legitimacy, and protecting the server; it employs network agents. This method can effectively thwart DDoS attacks by serving all legal requests, according to simulation data; here the Dispatcher plays a vital role in load balancing or sharing traffic to the free servers, but in the proposed system, there is no load balancing to the servers, monitoring the traffic and based on the analysis of the traffic sends alerts and sometimes ends up blocking the source node for a certain amount of time. The DDoS defence system, which includes attack detection and mitigation and might be implemented in the network domain, was developed by Fayaz [6] and demonstrated the elements and the purest process that a whole defence system should have; we intend to utilise it as one of the examples. To mitigate different DDoS attack combinations, one must dynamically detect them, design a responsive resource management algorithm that is 4–5 orders of magnitude faster than the state-of-the-art solvers at the time, deploy an industrial-grade SDN/NFV platform and experiment with 10 Gbps attack traffic.

The responsiveness, resource utilisation effectiveness, and multi-functionality of the above functions need to be revised for this. However, it sets the standard route for all DDoS defence systems after this. In the few years after the release of Bohatei, many DDoS defence systems based on software-defined networking and network virtualisation have come out one after the other. Dynamically creating and destroying the agents for a scalable and flexible system based on the density of traffic was developed by Bahman Rashidi [13]. Reference has been taken from this unit for creating and destroying the nodes in the NFV environment dynamically based on the requests and traffic flow.

3. Proposed Model

NFV is a technology that enables the virtualisation of network functions, allowing for deploying and managing network services on virtual machines or containers. However, the virtualised nature of NFV also makes it vulnerable to DDoS attacks, which can disrupt or disable network services. Preventing DDOS attacks on NFV requires a comprehensive defence strategy that includes techniques such as the filtering and diverting of traffic, alerts and alarms, and limiting the incoming traffic rate. The main aim is to stop malicious traffic before it enters the virtual server environment. Based on this, a model was proposed to protect the NFV environment from DDOS attacks. The proposed model will detect an attack in the early stages, send warnings to the source nodes, and then block the traffic for a particular time interval.

3.1 Proposed Algorithm Assumptions

Before algorithm initialisation, we assume a few parameters which may be used in the future. These parameters are:

1. Global Threshold value
2. Global Timer value
3. Local Timer value

4. The local Counter value (By default set to '0')

5. Local Limit value

Step 1: In the initial stage, the server accepts any number of requests until the global threshold value is crossed. The global timer (in seconds) indicates how long the server will perform steps 2 and 3. Only when the number of requests on the server exceeds the global threshold value does the server start its global timer and move to the next step.

Step 2: In this step, the initial scanning of nodes will be done. The server will identify the total requests sent by individual nodes during the Global Threshold break. After scanning, the server will set a limit for each node to the initial assumed Local Limit value. The server will calculate a local threshold (number of requests) value to each node such that each threshold value = 90% of the original number of requests sent by the node at the end of step 1. The server will set a counter value for each node to the initial assumed Local Counter value to track the number of times the node has crossed its local threshold. The server will also set up a local timer for each node to initially assume a Local Timer value to indicate the amount of time (in seconds) the node will be banned if it ever crosses the limit.

Step 3: The server will perform a DDoS prevention algorithm (Judge and Police) on each node that crosses their local threshold. The result of the Judge and Police algorithm can be either:

1. The server will detect a node sending too many requests (Or)

2. The server will slow down the number of requests sent from that node.

   - In the 2nd case, we move on.

   - In the 1st case, the server will increase its local counter value and reset the global timer. We then check if Local Counter crossed the Local Limit value; if yes, we orange list the node and warn neighbouring networks. After the limit is crossed, the server will stop accepting requests or, in other words, ban the node for its given time (the local timer variable). If the same node is orange listed for more than its Local Limit value, the server will stop accepting responses permanently. After the local timer reaches 0, it resets itself such that the value now will be twice as before. Also, the local timer will never reset to the original value. If the global timer reaches 0, then the server will reset the variables and move to step 1.

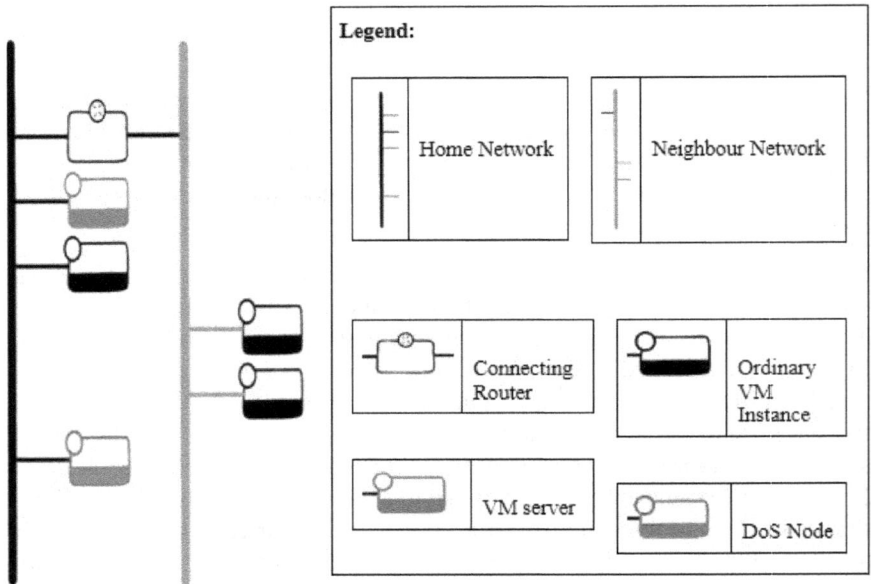

Figure 4. A NFV network scenario that defends against DDoS attacks on the target using multiple agents.

In Fig. 4, let's consider a scenario where the VM which is acting as a server, can only receive 1,111 requests before going down. We set this server's Global Threshold Value to 90% of the total number of requests it can receive

- $G\_th = (1111/100) * 90 \approx 1000$ requests per second
- Global Timer value: $G\_t = 200$ seconds
- Local Timer Value: $L\_t = 60$ seconds
- Local Counter value: $L\_c = 0$
- Local Limit value: $L\_lim = 4$

Step 1: Let's consider that the other VM instances start sending requests to the VM se. The server accepts any number of requests per second until the Global threshold is reached. Once the number of requests crosses the Global threshold value, which in this case it does because of the DoS VM, the Global Timer starts.

Step 2: Until local threshold values are calculated, the server stops accepting responses. Let's say the Dos VM is node A, the ordinary VM at the home network is node B and the ordinary VM at the neighbour network is Node C. Let's say the number of requests sent by Node A: $R\_A = 750$. The number of

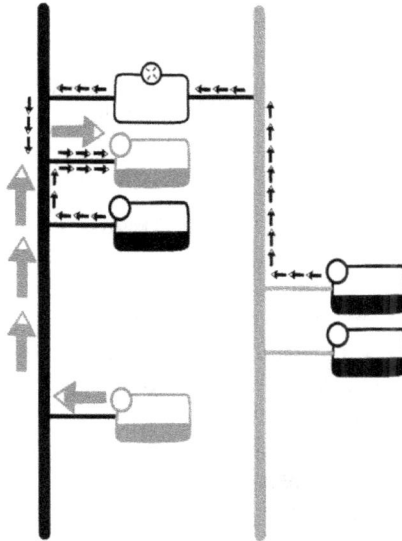

Figure 5. Detecting DDoS attack.

requests sent by Node B: $R\_B = 100$. The number of requests sent by Node C: $R\_C = 150$

We calculate the local variables for each node that sent a request at the time when T_g is crossed:

- L_th_A: $0.9 * (R\_A) = 0.9 * 750 = 675$
- L_th_B: $0.9 * (R\_B) = 0.9 * 100 = 90$
- L_th_C: $0.9 * (R\_C) = 0.9 * 150 = 135$

Now the server again starts accepting responses and we move to step 3.

Step 3: Node A again sends a huge number of requests but now since we have the local threshold value for each node, we check if $R\_A > L\_th\_A$. If yes, then we perform a judge and police algorithm. This tries to slow down the number of requests that are being sent, but fails as it is a DoS node so then we increase Local Counter, L_c by 1 and the Global Timer is reset to its initial value. We then check if the Local counter, $L\_lim < L\_c$. If yes, then the local timer starts, we stop accepting responses from Node A, orange list Node A and send a warning about Node A (it's IP address) to neighbour networks. It should be noted that all of this is happening while the global timer is running and we stop accepting any response from Node A until the Local Timer sets to 0. Once the Local Timer is set to 0, it resets to twice the initial value. This cycle will repeat until eventually Local Timer > Global Timer. At this point, the server will move again to step 1. If Node A gets orange listed more than the local limit value, we will stop accepting requests permanently. This is how we make a DoS node incapable of sending requests.

3.2 To Detect DDoS Attack

Traffic enters into the NFV environment and then its nodes where it is organised according to the sources. The threshold is calculated for requests per second after the current threshold crosses the global threshold value the detecting algorithm encounters. The algorithm first calculates the density of each node and sets up a local threshold value for each node. If a node crosses its local threshold value when sending a number of requests, then the police algorithm runs for that node and the counter is raised by 1. Now we have a limit variable declared local to each node which indicates the limit for the number of times the local threshold value is crossed. After each time a node crosses the local threshold, we check if (counter > limit). If yes, then we orange list that node. It gives a warning to the source that the traffic is more. Then after the node value crosses the limit it blocks the traffic for a certain amount of time and concludes it as the attack. Figure 6, shows the steps to detect a DDoS attack.

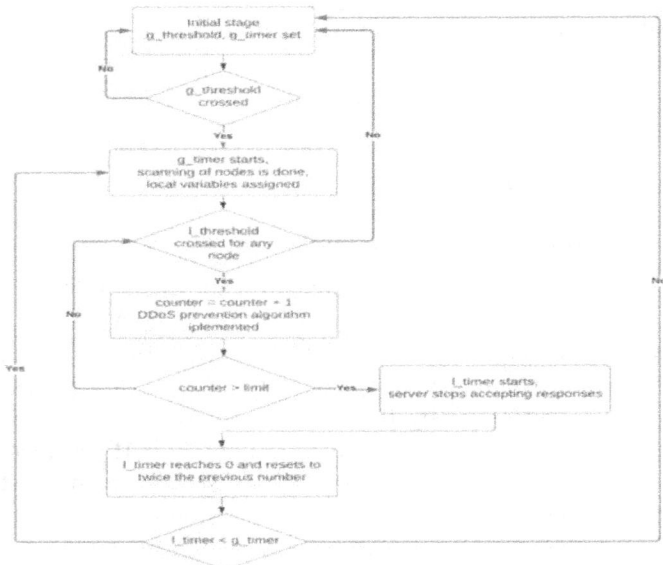

Figure 6. Flowchart shows the operations of the proposed method to detect DDoS attacks using NFV.

4. Implementation of DDoS Prevention System using NFV in Digital Twin Environment

A flowchart is a visual representation of a process or algorithm using different symbols and connectors to depict the sequence of steps. It is commonly used in computer programming, software engineering, and process documentation to illustrate the flow of control or the logical structure of a program. Flowcharts consist of various shapes or symbols that represent different actions or decisions. Following the flowchart from the start symbol, one can understand the sequence of actions and decision points to reach the desired outcome. The flowchart uses arrows or lines to connect the symbols and show the flow of execution from one step to another. Arrows indicate the direction and sequence of the process or decision-making. Flowcharts are an effective way to visualise complex processes, identify potential bottlenecks or errors, and communicate the logic of a program or system.

4.1 Digital Twin Environment to Detect and Prevent DDoS

Case 1 - Attacker node in the same Network:

The Attacker node/VM attacks our VM/node. The VM detects a DoS attack from this VM/node. Now the VM blocks the attacker node, and the VM sends the signal to the local router. The router sends a warning about this node to its directly connected neighbours.

Case 2 - Attacker node in a different network:

The VM detects a DoS attack from this VM/node, and the VM blocks the attacker node. The VM sends a signal to the local router about this node to its directly connected neighbours. Finally, the attacker network's local router turns off the attacker node.

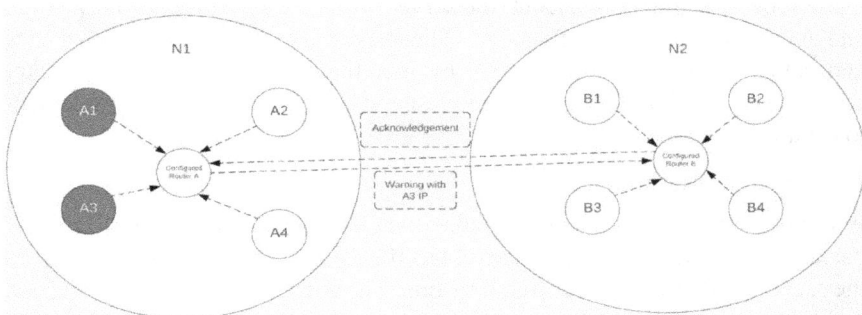

Figure 7. Network cluster diagram before DDoS detection mechanism in DT.

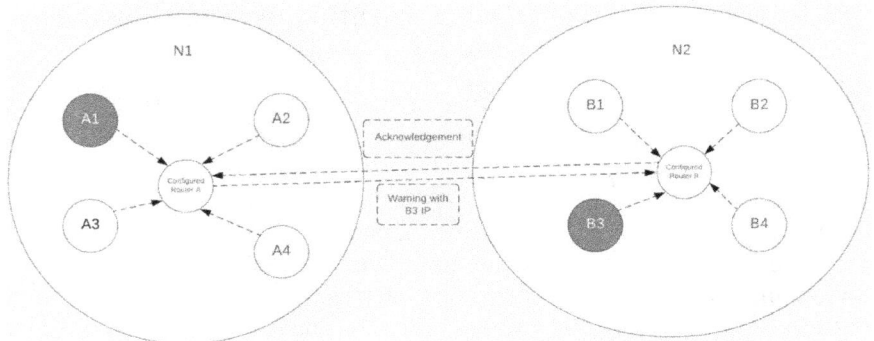

Figure 8. Network cluster diagram after the DDoS detection mechanism in DT.

4.2 The 3-way TCP Handshake Procedure

The TCP (Transmission Control Protocol) handshake is a three-step process to establish a connection between two devices in a TCP/IP network. It is essential to the TCP protocol suite and ensures reliable and orderly communication between the sender and receiver. Let's go through each step of the TCP handshake:

Step 1: SYN (Synchronise)

The initiating device, often called the client, sends a TCP packet with the SYN (synchronise) flag set to the receiving device, the server. The packet contains a randomly generated sequence number (ISN - Initial Sequence Number) to start the sequence numbering for data transmission. This step indicates the client's request to establish a connection and synchronise sequence numbers with the server.

Step 2: SYN-ACK (Synchronise-Acknowledge)

Upon receiving the SYN packet, the server sends a TCP packet with the SYN and ACK (acknowledge) flags set. The server also generates its own ISN and acknowledges the client's ISN by incrementing it by one. This packet confirms the receipt of the client's request and indicates the server's agreement to establish a connection.

Step 3: ACK (Acknowledge)

In the final step of the TCP handshake, the client sends an acknowledgement packet back to the server. The packet has the ACK flag set and acknowledges the server's ISN by incrementing it by one. The connection is established now, and both devices are ready to exchange data [5]. After the TCP handshake is completed, data transmission can begin. The sequence numbers established during the handshake are used to ensure that packets are delivered in the

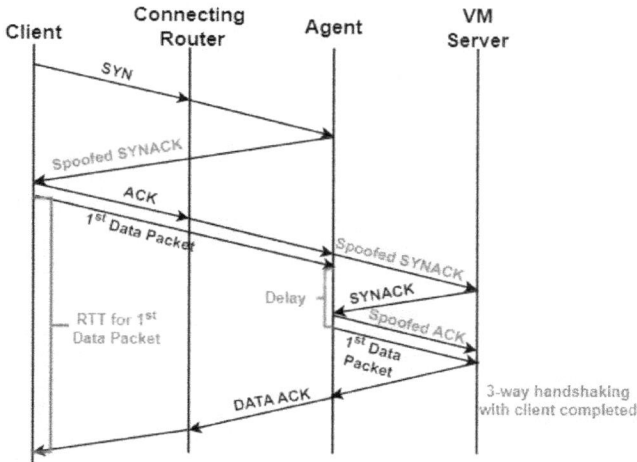

Figure 9. Spoofed TCP 3-way handshaking with a legitimate client and an attack.

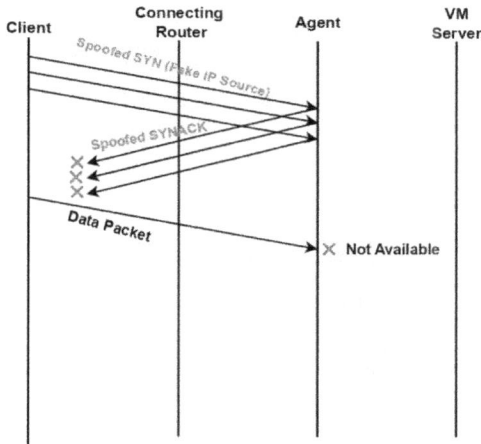

Figure 10. Attack scenario where a DDoS attack is prevented.

correct order and to detect and recover from any packet loss or errors during transmission. When the communication is finished, a similar process called the TCP connection termination or teardown occurs to close the connection gracefully.

OpenStack is an open-source cloud computing platform that provides software tools for building and managing public and private clouds. It allows users to control and allocate computing, storage, and networking resources through a centralised dashboard or API. OpenStack is highly flexible and scalable, making it suitable for building public, private, and hybrid cloud environments. It promotes interoperability and avoids vendor lock-in by adhering to open standards and providing an ecosystem of compatible solutions. OpenStack is used by organisations of all sizes, including enterprises, service providers, research institutions, and government agencies, to create and manage their cloud infrastructure [9].

5. Test Evaluation

This section primarily describes the evaluation of various testing scenarios of our proposed framework and discusses the obtained results. For testing purposes, the authors in this paper simulate the DDoS attack but do not suppose any attacks in the network organizations, especially in SDN environments.

5.1 Test Description

For our proposed framework, we analysed three test scenarios. The test scenarios of our proposed framework are mentioned in Table 2. We have observed that the attack should be mitigated to its source domains from the three test scenarios. The proposed framework is verified under three server domains. For practicality, mitigating the DOS attack in thousand nodes in an SDN environment should require a separate testbed and need a separate network simulator.

Table 1. Proposed model test scenarios.

Test scenarios	Description	Network parameters considered	Host ping arguments
I	DDoS attack with spoofed IP address simulated with one malign host and two benign hosts	Round Trip Time, Packet loss, and DDoS mitigation time	TCP-SYN, FLOOD
II	DDoS attack simulated with two malign hosts	Round Trip Time, Packet loss, and DDoS mitigation time	TCP-SYN, FLOOD
III	DDoS attack with encapsulated packets size with one malign host and two benign hosts	Round Trip Time, Packet loss, and DDoS mitigation time	TCP-SYN, FLOOD

Table 2. The different domain servers' performance without DDoS Attacks in DT.

Test case no	Server1		Server2		Server3	
	Average RTT (ms)	Packet Loss (%)	Average RTT (ms)	Packet Loss (%)	Average RTT (ms)	Packet Loss (%)
1	0.645	0	0.452	0	0.748	0
2	0.421	0	0.653	0	0.542	0
3	0.754	0	1.231	0	0.245	0
4	0.341	0	0.546	0	1.365	0
Average	0.540	0	0.720	0	0.725	0

5.2 Discussion on Testing Scenarios'

In this subsection, we have presented a detailed discussion and practicality of our proposed framework. During the transmission of packets from the source to the destination within the network, the switch provides a timestamp and a number of packets transmitted in the network. Table 3 provides quality of service (QoS) metrics where there is no DDoS attack observed in three domain servers. Table 4 shows the three testing scenarios of DDoS attacks. The host IP address for DDoS attack mitigation is 10.1.60.43, and the destination machine is "DESKTOP-UQ61BU1". The QoS metrics for the DDoS attack mitigation time were collected from the BurpSuite [19]. The remaining parameters, such as average round trip time (RTT) and percentage of packet loss retrieved from the ping command of host machines.

Table 3. Performance results in three test scenarios'.

Test no.	Test scenario I			Test scenario II			Test scenario III		
	DDOS Mitigation Time (s)	Average RTT (ms)	Packet Loss (%)	DDOS Mitigation Time (s)	Average RTT (ms)	Packet Loss (%)	DDOS Mitigation Time (s)	Average RTT (ms)	Packet Loss (%)
1	2	0.621	0	2	0.321	0	3	0.464	0
2	4	0.487	0	3	0.236	0	2	0.491	0
3	5	0.654	0	2	0.471	0	2	0.482	0
4	6	0.753	0	3	0.638	0	2	0.364	0
Avg	4.25	0.628	0	2.5	0.416	0	2.25	0.450	0

Table 4. The results after comparing all scenarios of our proposed method.

Scenario	DDOS Mitigation Time (s)	Average RTT (ms)	Packet Loss (%)
Normal behavior	---	0.661	0
Scenario I	4.25	0.628	0
Scenario II	2.5	0.416	0
Scenario III	2.25	0.450	0
Avg	2.25	0.538	0

6. Conclusion and Future Works

The early detection and prevention of DDoS attacks in NFV environments is a critical area of research and development. As NFV continues to be adopted in various network architectures, the need for robust security measures becomes increasingly important. This project aims to safeguard NFV environments against the disruptive and damaging effects of DDoS attacks by focusing on early detection, proactive monitoring, and effective mitigation strategies. The project on early detection and prevention of DDoS attacks in Network Function Virtualisation (NFV) environments is of utmost importance in ensuring the security and availability of networks. By focusing on early detection and prevention, the project aims to mitigate the disruptive effects of DDoS attacks and maintain the optimal performance of NFV systems. Furthermore, integrating Software-Defined Networking (SDN) with NFV can provide a centralised and programmable control plane, enabling rapid security policy deployment and real-time traffic redirection. This integration enhances the agility and responsiveness of DDoS defence mechanisms.

Ultimately, the project on early detection and prevention of DDoS attacks in NFV contributes to developing robust and proactive security strategies for NFV environments. By staying ahead of evolving threats and leveraging innovative technologies, the project helps safeguard critical networks and services from the disruptive consequences of DDoS attacks. Ultimately, the early detection and prevention of DDoS attacks in NFV environments serve as a vital defence mechanism in the face of evolving and sophisticated cyber threats. Through continued research, development, and collaboration, we can create a more resilient and secure network ecosystem to benefit organisations and end-users. The findings and outcomes of this project have the potential to significantly enhance the security posture of NFV environments, ensuring the availability, integrity, and confidentiality of network services. By staying at the forefront of DDoS attack detection and prevention methods, this project contributes to the ongoing efforts to combat cyber threats and protect critical network infrastructures.

Future Works

In the future, detecting and preventing Distributed Denial of Service (DDoS) attacks in Network Function Virtualisation (NFV) environments is likely to focus on several key areas. Here are some potential directions for future work in this field:

- *Machine Learning and AI-Based Approaches*: The use of machine learning and artificial intelligence techniques can enhance DDoS detection and prevention in NFV. By analysing network traffic patterns and behaviour, these algorithms can learn to identify anomalous activities associated with DDoS attacks, enabling early detection and rapid response.

- *Behavioural Analysis*: Future solutions may employ advanced behavioural analysis techniques to identify patterns and characteristics of normal network behaviour. By continuously monitoring and analysing network traffic, anomalies associated with DDoS attacks can be detected more accurately, even in complex NFV environments.

- *Dynamic Resource Allocation*: NFV allows for dynamic allocation of network resources. Future work can focus on developing intelligent resource allocation mechanisms that can rapidly scale up or down the network capacity based on the current DDoS attack situation. This flexibility enables the mitigation of attacks by redirecting traffic, provisioning additional resources, or utilising specialised security functions.

- *Collaborative Defense Mechanisms*: In the future, DDoS in NFV environments may involve collaborative efforts between multiple network entities. Information sharing among network operators, service providers, and security platforms can enable the early detection and mitigation of attacks by leveraging collective intelligence and coordinated response strategies.

- *SDN Integration*: Software-Defined Networking (SDN) and NFV often go hand in hand. Future research can focus on integrating SDN and NFV to enhance DDoS defense capabilities. SDN's centralised control and programmability can facilitate the rapid deployment of security policies and enable real-time traffic redirection to mitigate attacks more effectively.

- *Advanced Mitigation Techniques*: In the future, NFV environments may adopt more advanced mitigation techniques to combat DDoS attacks. This could include the use of in-network scrubbing centers, traffic filtering and rate limiting, traffic diversion, or even leveraging cloud-based resources for scalable DDoS protection.

References

[1] Balarezo, J.F., Wang, S., Gomez Chavez, K. Al-Hourani, A. and Kandeepan, S. 2022. A survey on DoS/DDoS attacks mathematical modeling for traditional, SDN and virtual networks. Engineering Science and Technology an International Journal 31: 101065.

[2] Alnaim, A.K., Alwakeel, A.M. and Fernandez, E.B. 2022. Towards a security reference architecture for NFV. Sensors 2022 May 14; 22(10): 3750.

[3] Rashidi, B., Fung, C. and Rahman, M. 2018. A Scalable and Flexible DDoS Mitigation System Using Network Function Virtualization, IEEE.

[4] Girma, A., Garuba, M., Li, J. and Liu, C. 2015. Analysis of DDoS attacks and an introduction of a hybrid statistical model to detect DDoS attacks on cloud computing environment. pp. 212–217. *In*: International Conference on Information Technology - New Generations, IEEE.

[5] Li, H. and Wang, L. 2018. Online orchestration of cooperative defense against ddos attacks for 5g mec. pp. 1–6. *In*: 2018 IEEE Wireless Communications and Networking Conference (WCNC), IEEE, Barcelona, Spain.

[6] Fayaz, S.K., Tobioka, Y., Sekar, V. and Bailey, M. 2015. Bohatei: Flexible and elastic {DDoS} defense. pp. 817–832. *In*: 24th USENIX Security Symposium (USENIX Security 15), Washington, D.C., USA.

[7] Li, Q., Huang, H., Li, R. Lv, J., Yuan, Z., Ma, L., Han,Y. and Jiang, Y.A. 2023. Comprehensive Survey on Ddos Defense Systems: New Trends and Challenges. Available at SSRN 4363418.

[8] Hoque, N., Bhattacharyya, D.K. and Kalita, J.K. 2015. Botnet in DDoS attacks: trends and challenges. IEEE Communications Surveys & Tutorials 2015 Jul 16; 17(4): 2242–70.

[9] Bojjagani, S., Brabin, D.R. and Saravanan, K. 2022. Early DDoS detection and prevention with traced-back blocking in SDN environment. Intelligent Automation & Soft Computing. 2022 Nov 1; 34(2).

[10] Jakaria, A.H., Yang, Rashidi, W.B., Fung, C. and Rahman, M.A. 2016. Vfence: A defense against distributed denial of service attacks using network function virtualization. pp. 431–436. In 2016 IEEE 40th Annual Computer Software and Applications Conference (COMPSAC) 2016 Jun 10 (Vol. 2). IEEE.

[11] Singh, A.K., Jaiswal, R.K. Abdukodir, K. and Muthanna, A. 2020. ARDefense: DDoS detection and prevention using NFV and SDN. pp. 236–241. In 2020 12th International Congress on Ultra Modern Telecommunications and Control Systems and Workshops (ICUMT). IEEE.

[12] Rashidi, B. and Fung, C. 2016. CoFence: A collaborative DDoS defense using network function virtualization. pp. 160–166. In 2016 12th International Conference on Network and Service Management (CNSM) 2016 Oct 31. IEEE.

[13] Rashidi, B., Fung, C. and Rahman, M. 2018. A scalable and flexible DDoS mitigation system using network function virtualization. pp. 1–6. NOMS 2018 - 2018 IEEE/IFIP Network Operations and Management Symposium, Taipei, Taiwan, 2018, doi: 10.1109/NOMS.2018.8406314.

[14] OpenStack and NFV(Telicloud)-Asad Khan, Udemy.

[15] ETSI. Network Functions Virtualisation (NFV); Architectural Framework; ETSI: Sophia Antipolis, France, 2014.

[16] Bojjagani, S., Denslin Brabin, D.R. and Venkateswara Rao, P.V. 2020. Phishpreventer: a secure authentication protocol for prevention of phishing attacks in mobile environment with formal verification. Procedia Computer Science 171: 1110–1119.

[17] Bojjagani, S. and Sastry, V.N. 2017. VAPTAi: a threat model for vulnerability assessment and penetration testing of Android and iOS mobile banking apps. pp. 77–86. In 2017 IEEE 3rd International Conference on Collaboration and Internet Computing (CIC). IEEE.

[18] Bojjagani, S., Sastry, V.N., Chen, C-M. Kumari, S. and Khan, M.K. 2023. Systematic survey of mobile payments, protocols, and security infrastructure. Journal of Ambient Intelligence and Humanized Computing 14(1): 609–654.

[19] Online Available: https://portswigger.net/burp (Accessed: 2023-09-08).

CHAPTER 11

Security and Privacy Considerations in Blockchain-based IoT Systems

Anjali V. S.

1. Introduction

The Internet of Things (IoT) has witnessed remarkable and rapid growth in recent years, transforming the way we interact with technology and revolutionising various industries. IoT refers to the network of interconnected devices, sensors, and systems that can collect, exchange, and analyse data. From smart homes and wearable devices to industrial automation and smart cities, IoT has permeated numerous aspects of our lives.

This explosive growth of IoT brings with it significant implications for security and privacy [1]. As the number of interconnected devices continues to surge, the need to ensure the integrity and confidentiality of data becomes paramount. IoT devices are vulnerable to various security threats, including unauthorised access, data breaches, and manipulation. Compromised devices can not only disrupt operations but also compromise the privacy of individuals and organisations.

The significance of security and privacy in the IoT context cannot be overstated. The potential consequences of security breaches range from financial losses to physical harm, making it imperative to address these challenges proactively. The Internet of Things (IoT) has witnessed remarkable

SRM University AP.
Email: anjalisajayan@gmail.com

and rapid growth in recent years, transforming the way we interact with technology and revolutionising various industries [2]. IoT refers to the network of interconnected devices, sensors, and systems that can collect, exchange, and analyse data. From smart homes and wearable devices to industrial automation and smart cities, IoT has permeated numerous aspects of our lives.

This explosive growth of IoT brings with it significant implications for security and privacy [3]. As the number of interconnected devices continues to surge, the need to ensure the integrity and confidentiality of data becomes paramount. IoT devices are vulnerable to various security threats, including unauthorised access, data breaches, and manipulation. Compromised devices can not only disrupt operations but also compromise the privacy of individuals and organisations [4].

The significance of security and privacy in the IoT context cannot be overstated. The potential consequences of security breaches range from financial losses to physical harm, making it imperative to address these challenges proactively. Furthermore, with IoT devices collecting vast amounts of personal data, protecting privacy becomes a critical concern. Individuals should have control over their personal information and be confident that it is not being exploited or misused.

The interconnected nature of IoT systems amplifies the complexity of security and privacy challenges. Traditional security approaches often fall short of providing adequate protection for IoT devices due to factors such as resource constraints, heterogeneity, and scalability issues [5]. Similarly, privacy concerns arise from the continuous data collection and aggregation, necessitating robust mechanisms to safeguard personal information.

Recognising these challenges, researchers, industry practitioners, and policymakers have turned to blockchain technology as a potential solution to enhance security and privacy in IoT systems. Blockchain, known for its decentralised and immutable nature, offers the potential to create a secure and trustworthy infrastructure for IoT. By leveraging cryptographic techniques, consensus algorithms, and smart contracts, blockchain can provide a secure data exchange, tamper-resistant storage, and fine-grained access control, bolstering the overall security and privacy of IoT systems.

2. Security Challenges in IoT Systems

IoT systems' interconnectedness and the wide variety of devices they use provide several security problems. It is essential to comprehend these issues in order to create security solutions that work [6].

Here are a few typical IoT system security issues:

Device Vulnerabilities: IoT devices often have limited computing power and resources, which can make them vulnerable to various attacks. Weak or default passwords, outdated firmware, and lack of secure communication protocols can be exploited by attackers to gain unauthorised access or control over the devices.

Data Integrity and Authenticity: Ensuring the integrity and authenticity of data transmitted and stored by IoT devices is essential. Data tampering or manipulation can lead to false information, compromised decision-making, and potentially dangerous actions. Verifying the source and integrity of data in a decentralised IoT environment can be challenging.

Network Security: IoT devices rely on network communication for data exchange, making network security a critical concern. Inadequate network security measures can expose IoT devices and their data to various threats, including eavesdropping, Man-in-the-Middle attacks, and denial-of-service (DoS) attacks.

Lack of Standardisation: The lack of standardised security protocols and frameworks across IoT devices and platforms makes it difficult to establish consistent security practices. Incompatibilities and inconsistencies can lead to vulnerabilities and gaps in the security of interconnected devices.

Scalability and Complexity: IoT ecosystems can involve a vast number of devices with different functionalities and communication protocols. Managing and securing such a complex and dynamic environment poses significant scalability challenges, requiring robust security mechanisms that can handle the increasing number of devices and interactions.

Privacy Concerns: IoT devices often collect sensitive personal data, raising privacy concerns. Unauthorised access to this data can result in privacy breaches and identity theft. Additionally, the constant monitoring and tracking capabilities of IoT devices raise questions about individual privacy and consent.

Legacy System Integration: Many IoT implementations involve integrating new technologies with existing legacy systems. Legacy systems may lack proper security measures or be incompatible with modern security protocols, creating potential entry points for attackers.

Supply Chain Risks: The global supply chain for IoT devices involves multiple vendors and manufacturers, making it vulnerable to security risks.

Compromised devices introduced during the manufacturing process or supply chain can have built-in vulnerabilities that can be exploited.

In order to address these security issues, a multifaceted strategy that considers data encryption, network security, secure communication protocols, device security, and access control measures is necessary. To maintain the dependability and durability of IoT systems as the IoT environment changes, security must be prioritised at every level of design, development, deployment, and operation [7].

3. Blockchain Technology for IoT Security

In order to improve security in IoT networks, blockchain technology has emerged as a possible alternative. It offers various advantages for protecting IoT devices and data by utilising the inherent properties of blockchain, such as decentralisation, immutability, and consensus procedures.

The following are some crucial ways that blockchain technology might improve IoT security:

Decentralisation: Blockchain operates in a decentralised manner, eliminating the need for a central authority or trusted third party. This decentralised architecture enhances the security of IoT systems by removing single points of failure and reducing the risk of unauthorised access or control. Each participant in the blockchain network maintains a copy of the distributed ledger, ensuring transparency and integrity.

Immutability: Blockchain's immutability ensures that once data is recorded on the blockchain, it becomes tamper-resistant. This property makes it highly suitable for ensuring the integrity and authenticity of data generated by IoT devices. Any attempt to alter or tamper with the data recorded on the blockchain would require the consensus of a majority of network participants, making it extremely difficult for attackers to modify or manipulate data without detection.

Data Integrity and Auditability: The transparent nature of blockchain enables the verification and auditability of data transactions. Every transaction recorded on the blockchain is time-stamped, creating an immutable audit trail. This feature enhances data integrity, providing a robust mechanism for detecting any unauthorised changes to IoT device data.

Secure Communication and Identity Management: Blockchain technology can facilitate secure communication between IoT devices by establishing a decentralised and encrypted peer-to-peer network. Smart contracts deployed on the blockchain can enable secure device authentication, access control, and data sharing. This ensures that only authorised devices can participate in the network and access specific data, reducing the risk of unauthorised access or malicious activities [8].

Consensus Mechanisms: Blockchain networks rely on consensus mechanisms to validate and agree upon the state of the ledger. Consensus algorithms such as Proof of Work (PoW) or Proof of Stake (PoS) ensure the validity and consistency of data across the network. These mechanisms provide additional security by requiring network participants to solve cryptographic puzzles or stake their tokens, making it computationally expensive and economically unfeasible for attackers to manipulate the blockchain.

Tamper-Resistant Device Updates: Blockchain technology can facilitate secure and tamper-resistant firmware updates for IoT devices. By using blockchain, manufacturers can digitally sign firmware updates and record them on the blockchain. This ensures the authenticity and integrity of the updates, mitigating the risk of unauthorised or malicious firmware modifications.

Trust and Transparency: The use of blockchain technology enhances trust and transparency in IoT systems. Participants in the blockchain network can have increased confidence in the integrity and security of the data exchanged, as they can independently verify and audit the transactions recorded on the blockchain.

Even though blockchain technology has several benefits for IoT security, it is crucial to carefully weigh the trade-offs. IoT installations may face difficulties due to issues with scalability, resource limitations, and the energy consumption of some blockchain implementations. To balance security and operational efficiency in IoT systems, it is crucial to choose the right blockchain designs, consensus methods, and optimisation approaches.

IoT systems may achieve better security, data integrity, and trustworthy interactions between devices by utilising blockchain technology. Additional research and development will enable the technology to reach its full potential for protecting the IoT environment, which is quickly developing.

4. Privacy Considerations in Blockchain-based IoT

Privacy considerations play a crucial role in ensuring the responsible and ethical deployment of blockchain-based IoT (Internet of Things) systems. While blockchain technology offers advantages such as transparency and immutability, it also raises concerns about the confidentiality and privacy of sensitive data.

Here are key privacy considerations in blockchain-based IoT:

Data Leakage: In IoT systems, devices generate and exchange vast amounts of data. Blockchain's transparent nature poses challenges in protecting sensitive data from unauthorised access or exposure. Without appropriate measures, sensitive information such as personal identifiers, location data, and usage patterns could be exposed to unauthorised parties.

User Identification: Blockchain transactions are typically pseudonymous, using cryptographic addresses instead of real-world identities. However, it is crucial to recognise that de-anonymisation techniques can potentially link these addresses to real individuals. This raises concerns about user identification and the potential loss of privacy when linking IoT device activities to specific users.

Data Ownership and Consent: In blockchain-based IoT, data ownership and consent are critical aspects. IoT devices often collect data from various sources, including individuals. Ensuring that individuals have control over their data, understand how it is being used, and provide informed consent is crucial for maintaining privacy. Smart contracts and permissioned access mechanisms can be utilised to address these concerns.

Selective Disclosure: Not all data collected by IoT devices needs to be shared on the blockchain. Selective disclosure mechanisms can be employed to share only necessary and non-sensitive data while keeping sensitive information off-chain. This approach helps protect privacy by minimising the exposure of personal and confidential data.

Anonymity and Privacy Trade-offs: Balancing anonymity and privacy with transparency is a challenge in blockchain-based systems. While blockchain's transparency enhances trust and auditability, it can compromise individual privacy. Striking the right balance between transparency and privacy is important, considering the specific use cases, regulatory requirements, and user expectations.

Regulatory Compliance: Compliance with privacy regulations, such as the General Data Protection Regulation (GDPR), becomes crucial in blockchain-based IoT systems. Organiaations must ensure that their data handling practices align with the principles and requirements outlined by relevant privacy regulations to protect user rights and maintain legal compliance.

Secure Access Control: Strong access control mechanisms are vital for protecting privacy in blockchain-based IoT. Implementing granular permission models, where individuals have control over their data and can grant selective access to trusted entities, which helps prevent unauthorised access and preserves privacy.

Privacy by Design: Incorporating privacy considerations from the early stages of system design is essential. By following privacy-by-design principles, organisations can proactively identify and mitigate privacy risks, establish privacy policies, and ensure that privacy-enhancing techniques are integrated into the architecture and operation of blockchain-based IoT systems.

It is necessary to take a comprehensive strategy that integrates technological, legal, and ethical factors in order to address privacy issues in blockchain-based IoT systems. Building trust and ensuring the responsible use of IoT data inside blockchain ecosystems requires striking the proper balance between openness, data protection, user control, and regulatory compliance. To encourage privacy-preserving IoT deployments, businesses must view privacy as a core necessity and implement privacy-enhancing strategies and best practices.

5. Security Mechanisms in Blockchain-based IoT

IoT (Internet of Things) systems built on blockchain use a variety of security measures to guarantee the reliability, confidentiality, and integrity of the network. To safeguard IoT devices, data, and interactions, these systems make use of the distinctive features of blockchain technology [9].

The following are some essential security measures frequently applied in blockchain-based IoT:

Encryption: Encryption is a fundamental security mechanism used to protect the confidentiality of data transmitted and stored within IoT systems. End-to-end encryption can be employed to secure communication channels between IoT devices, ensuring that only authorised recipients can access and decipher the data.

Digital Signatures: Digital signatures play a crucial role in verifying the authenticity and integrity of data in a blockchain-based IoT system. Utilising asymmetric cryptographic methods, digital signatures offer a way to confirm the sender's identity and guarantee that the data was not altered during transmission.

Secure Smart Contracts: Smart contracts are self-executing contracts that run on the blockchain. They define the rules and conditions for interactions between IoT devices. By ensuring the security and correctness of smart contracts through code reviews, vulnerability assessments, and secure coding practices, the potential for malicious actions or vulnerabilities can be minimised.

Access Control: In order to manage access to IoT devices and data inside a blockchain network and to govern permissions, access control techniques are essential. Attribute-based access control (ABAC) or Role-based access control (RBAC) can be employed to define and enforce fine-grained access policies, ensuring that only authorised entities can interact with specific devices or access certain data.

Consensus Algorithms: Consensus algorithms are essential for maintaining the blockchain's integrity and security. They make it possible for network users to concur on the legitimacy of transactions and the sequence in which they are recorded on the blockchain. Proof of Work (PoW) and Proof of Stake (PoS) consensus algorithms assist in stopping fraudulent behaviours like double-spending and blockchain hacking.

Immutable Ledger: The immutability of the blockchain ensures that once data is recorded, it cannot be altered or deleted without consensus from the network participants. This property makes the blockchain a secure and tamper-resistant ledger, reducing the risk of unauthorised modifications or data tampering within the IoT system.

Auditing and Monitoring: Continuous auditing and monitoring mechanisms are essential for detecting and responding to security incidents in blockchain-based IoT systems. Real-time monitoring of device activities, network traffic analysis, and anomaly detection techniques help identify suspicious behaviour and potential security breaches. Furthermore, audit trails recorded on the blockchain provide a transparent and verifiable history of transactions, aiding in forensic analysis and incident response.

Secure Identity Management: Secure identity management is critical in blockchain-based IoT to ensure that only authorised and authenticated devices participate in the network. Public key infrastructure (PKI) and decentralised identity solutions can be utilised to manage device identities, authenticate communication, and prevent unauthorised access to the network.

By incorporating these security measures, blockchain-based IoT systems may better protect the interactions, data, and devices, improving the network's overall security posture. To achieve an ideal and successful security implementation, it is crucial to consider the unique requirements, limitations, and trade-offs of each mechanism in the context of the IoT ecosystem.

6. Privacy-preserving Techniques in Blockchain-based IoT

While utilising the advantages of blockchain technology, privacy-preserving strategies in blockchain-based IoT (Internet of Things) systems seek to safeguard the confidentiality and privacy of sensitive data. These methods allow people to limit the disclosure of sensitive information to approved parties while giving them control over their personal information.

The following are some frequently used techniques of IoT-based blockchain privacy-preserving techniques:

Pseudonymity: Blockchain networks typically use pseudonyms or cryptographic addresses instead of real-world identities. By utilising pseudonyms, the link between IoT device activities and specific individuals can be obfuscated, providing a certain level of privacy and anonymity.

Selective Disclosure: Selective disclosure mechanisms allow users to share specific pieces of data with authorised entities while keeping the rest of the information private. It ensures that only the necessary data is disclosed, preserving privacy and minimising exposure to sensitive information.

Encryption: Sensitive data which is kept on the blockchain or which is exchanged between IoT devices can be encrypted using encryption methods. Data is made secure and private by being encrypted using cryptographic techniques, which prevents access from unwanted parties.

Zero-Knowledge Proofs: Zero-knowledge proofs (ZKPs) enable the validation of a statement without revealing any additional information. In the context of blockchain-based IoT, ZKPs can be used to prove the correctness or validity of certain data or actions without exposing the underlying sensitive information.

Homomorphic Encryption: With homomorphic encryption, calculations may be made on encrypted data without having to first decode it. By ensuring that sensitive data is encrypted during processing, this method adds another degree of privacy protection to IoT systems based on blockchain technology.

Privacy-focused Blockchain Frameworks: Specialised blockchain frameworks, such as privacy-focused or privacy-enhanced blockchains, offer additional privacy features tailored to IoT applications [10]. These frameworks may employ advanced privacy-preserving techniques, such as ring signatures, zero-knowledge proofs, or mix networks, to ensure privacy in the context of IoT.

Off-Chain Data Storage: Not all data collected by IoT devices needs to be stored directly on the blockchain. Off-chain storage solutions can be utilised to store sensitive data separately, while only storing necessary metadata or cryptographic proofs on the blockchain. This approach helps protect privacy by minimising the exposure of sensitive information on the public blockchain.

Consent Management: Consent management mechanisms enable individuals to control the usage of their personal data within the blockchain-based IoT system. By providing explicit consent for data sharing and defining specific usage policies, individuals can maintain control over their data and ensure that it is used only for authorised purposes [11].

In IoT systems built on blockchain, the combination of various privacy-preserving methods offers a tiered approach to protecting sensitive data and privacy. To balance blockchain technology's advantages for openness and privacy, though, is vital. To provide a privacy-respecting and reliable system, the implementation of suitable privacy methods should take into account user expectations, user expectations, and the unique context of the IoT application.

7. Case Studies and Use Cases

Case studies and use cases demonstrate the practical implementation of blockchain-based IoT systems with enhanced security and privacy. They provide real-world examples of how organisations have leveraged blockchain technology to address the challenges and achieve the benefits of IoT while ensuring data integrity and privacy.

Here are a few notable case studies and use cases:

Supply Chain Management: Blockchain-based IoT systems have been employed to enhance traceability and transparency in supply chain management. By integrating IoT devices, such as sensors and RFID tags, with blockchain technology, organisations can securely track and verify the movement of goods, monitor conditions like temperature and humidity, and ensure the authenticity of products throughout the supply chain. This use case has been particularly prominent in the food and pharmaceutical industries.

Energy Grid Management: Blockchain-based IoT systems have been utilised for decentralised energy grid management. By integrating IoT devices, such as smart meters and energy sensors, with blockchain, energy consumption data can be securely recorded and validated on the blockchain. This enables transparent and efficient energy trading, optimised grid management, and accurate billing, while maintaining the privacy of consumers' energy usage information.

Healthcare Data Sharing: Blockchain-based IoT systems have been implemented to facilitate secure and privacy-preserving sharing of healthcare data. IoT devices, such as wearables and medical sensors, can collect patient data and are securely stored on the blockchain [12]. Healthcare providers and researchers can access the data with patient consent, ensuring privacy and data integrity while enabling collaborative research, personalised healthcare, and efficient data interoperability.

Smart Cities: Blockchain-based IoT systems have been applied to create smart city solutions. By integrating IoT devices like smart sensors, cameras, and infrastructure management systems with blockchain technology, cities can securely collect and share data related to traffic management, waste management, public safety, and environmental monitoring. This enables efficient resource allocation, real-time decision-making, and enhanced citizen participation while preserving data privacy and security.

Digital Identity Management: Blockchain-based IoT systems have been employed for decentralised and secure digital identity management [13]. By combining IoT devices with blockchain, individuals can have control over their digital identities and securely authenticate themselves in various applications. This use case eliminates the need for centralised identity providers, reduces the risk of identity theft, and ensures privacy and data integrity in identity management.

These case studies and use cases highlight the versatility and benefits of integrating blockchain technology with IoT systems. They demonstrate how blockchain increases security, data integrity, transparency, and privacy in various industries and domains [14].

To get the best outcomes, organisations must carefully assess the unique needs and context of their IoT deployments. Each implementation, after all, has its own concerns and obstacles.

8. Future Directions and Challenges

8.1 Future Directions

Scalability: Enhancing scalability is one of the main future initiatives for blockchain-based IoT. Blockchain networks must be able to handle an increasing volume of transactions and data without jeopardising speed as the number of IoT devices increases [15]. To meet this difficulty, research and development efforts are concentrated on scaling solutions like sharding, sidechains, and off-chain processing.

Interoperability: Achieving interoperability between different blockchain platforms and IoT devices is crucial for seamless integration and widespread adoption. Future directions involve the development of standards and protocols that enable cross-platform communication and interoperability, allowing diverse IoT devices and blockchain networks to interact efficiently.

Privacy Enhancements: Future advancements aim to enhance privacy in blockchain-based IoT systems. This includes the development of more sophisticated privacy-preserving techniques, such as advanced zero-knowledge proofs, secure multiparty computation, and privacy-focused consensus algorithms. Innovations in privacy-enhancing cryptography and decentralised identity management systems will also contribute to strengthening privacy in the IoT ecosystem.

Integration with AI and Machine Learning: The integration of artificial intelligence (AI) and machine learning (ML) with blockchain-based IoT systems holds immense potential. AI and ML algorithms can leverage the vast amount of data collected by IoT devices to derive valuable insights, optimise decision-making, and enhance the efficiency of IoT operations. Blockchain can ensure the integrity and immutability of AI and ML models and provide transparent auditing of their outputs.

8.2 Challenges

Here are some challenges related to Blockchain-based IoT systems

Energy Efficiency: Blockchain-based IoT systems can consume significant amounts of energy due to the computational requirements of consensus mechanisms. Overcoming the energy efficiency challenge is essential to make blockchain-based IoT systems sustainable [16]. To decrease energy use and environmental effect, researchers are investigating innovative consensus methods including proof-of-stake (PoS) and energy-efficient consensus algorithms.

Governance and Regulation: The decentralised nature of blockchain-based IoT raises challenges related to governance and regulation. Determining legal frameworks, establishing standards, and addressing liability and accountability issues in decentralised environments are areas that require further attention. Collaboration between industry stakeholders, policymakers, and regulatory bodies is crucial to developing appropriate governance models and regulations that foster innovation while safeguarding privacy and security.

Security Considerations: While blockchain provides enhanced security for IoT systems, new security challenges may emerge as the technology evolves. The potential vulnerabilities in smart contracts, consensus algorithms, and blockchain implementations must be continually addressed through robust security practices, code audits, and penetration testing. Staying vigilant against emerging threats, such as quantum computing, is also important.

User Experience and Adoption: Usability and user experience are critical for the widespread use of IoT technologies powered by blockchain. Simplifying user interfaces, integrating seamless authentication mechanisms, and educating end-users about the benefits and functionalities of blockchain-based IoT are necessary to foster acceptance and encourage adoption.

Addressing these future directions and challenges will require collaborative efforts from researchers, industry experts, policymakers, and technology providers [17]. Continued research and innovation, combined with practical implementations and regulatory support, will pave the way for the secure, privacy-preserving, and scalable integration of blockchain technology with IoT systems.

9. Conclusion

The incorporation of blockchain technology with IoT systems offers significant opportunities to enhance security, privacy, and efficiency. The accelerated growth of the Internet of Things has brought forth unique challenges in security, such as device vulnerabilities, data integrity concerns, and network attacks [18]. Blockchain technology offers inherent features such as decentralisation, immutability, and consensus mechanisms that address these challenges and provide a foundation for secure IoT systems.

Furthermore, privacy considerations in the blockchain-based IoT systems are crucial to safeguard sensitive data and ensure individuals' control over their personal information. Techniques such as pseudonymity, selective disclosure, encryption, and privacy-focused blockchain frameworks enable the preservation of privacy while maintaining transparency and data integrity.

Real-world use cases and case studies have demonstrated the practical implementation of blockchain-based IoT systems in various domains, including supply chain management, energy grid management, healthcare data sharing, smart cities, and digital identity management [19]. These examples showcase blockchain technology's advantages for improving security, transparency, and privacy while enabling efficient and trusted interactions in IoT ecosystems.

Looking ahead, future directions for blockchain-based IoT systems include addressing scalability challenges, promoting interoperability, enhancing privacy techniques, and integrating AI and machine learning [20].

Overcoming challenges related to energy efficiency, governance, security considerations, and user experience will be crucial for the widespread adoption and success of blockchain-based IoT systems.

By combining the strengths of blockchain and IoT, we can build a more secure, resilient, and privacy-preserving infrastructure for the ever-expanding interconnected world. Continued research, collaboration, and innovation among researchers, industry practitioners, and policymakers will drive the advancement of blockchain-based IoT systems and pave the way for a trustworthy and efficient IoT ecosystem.

References

[1] Christidis, K. and Devetsikiotis, M. 2016. Blockchains and Smart Contracts for the Internet of Things. IEEE Access 4: 2292–2303.

[2] Dorri, A., Kanhere, S.S. and Jurdak, R. 2017. Blockchain for IoT security and privacy: the case study of a smart home. pp. 618–623. In Proceedings of the IEEE International Conference on Pervasive Computing and Communications Workshops (PerCom Workshops). IEEE.

[3] Zheng, Z., Xie, S., Dai, H.N., Chen, X. and Wang, H. 2018. Blockchain challenges and opportunities: A survey. International Journal of Web and Grid Services 14(4): 352–375.

[4] Conti, M. and Dehghantanha, A. 2019. Security, privacy and trust in Internet of Things: The road ahead. Computer Networks 148: 221–222.

[5] Li, X., Jiang, P., Chen, T., Luo, X. and Wen, Q. 2018. A survey on the security of blockchain systems. Future Generation Computer Systems 82: 395–411.

[6] Li, M., Yu, S., Zheng, Y. and Ren, K. 2019. Blockchain-based privacy-preserving and secure data sharing for IoT. IEEE Network 33(5): 220–226.

[7] Moser, M. and van der Heijden, R.W. 2018. Secure data sharing in decentralized IoT environments using blockchain and smart contracts. pp. 80–85. In Proceedings of the IEEE 2nd International Workshop on Secure Smart Cities (SmartCitySec). IEEE.

[8] Kouicem, D.E., Kamoun, F. and Bouabdallah, A. 2019. A survey on blockchain-based IoT system: Architecture, applications, and future trends. IEEE Access 7: 114587–114602.

[9] Shrestha, R. and Mahmood, A.N. 2019. A systematic review of blockchain for secure and privacy-preserving storage and sharing of IoT data. IEEE Internet of Things Journal 6(5): 8291–8303.

[10] Zeng, H., Sheng, Z., Qin, Y., Li, X. and Wu, J. 2019. A survey on consensus mechanisms and mining strategy management in blockchain networks. IEEE Access 7: 22328–22370.

[11] Fan, K., Chen, J., Wu, P., Li, M. and Deng, R.H. 2018. A blockchain-based privacy-preserving data sharing scheme for IoT devices. Future Generation Computer Systems 86: 1383–1393.

[12] Dagher, G.G., Mohler, J., Milojkovic, M., Marella, P.B. and Al-Bassam, M. 2018. Ancile: Privacy-preserving framework for access control and interoperability of electronic health records using blockchain technology. Sustainable Cities and Society 39: 283–297.

[13] Boudguiga, A., Meddeb, A. and Tabbane, S. 2020. Blockchain and IoT integration: Challenges and opportunities. Future Internet 12(1): 7.

[14] Cao, Y., Hu, Y., Liu, Z., Niyato, D., Wang, P. and Wen, Y. 2019. Privacy-preserving and scalable blockchain-based architecture for sharing electronic health records. IEEE Transactions on Industrial Informatics 15(1): 487–498.

[15] Radanović, G., Likić, R., Matijasevic, M., Radenković, B. and Krčo, S. 2018. A blockchain-based access control framework for the Internet of Things. Sensors 18(10): 3318.

[16] Liu, J., Liang, X., Chen, Y., Li, T. and Chen, J. 2019. Blockchain-based data integrity service framework for IoT data. Future Generation Computer Systems 100: 132–144.

[17] Li, Z., Liang, H., Y. Zuo, Y. and Wu, W. 2020. A novel secure and efficient authentication scheme for Internet of Things leveraging blockchain technology. IEEE Internet of Things Journal 7(7): 5784–5796.

[18] Li, Q., Cao, J., Wang, C., Qin, C. and Li, M. 2019. BIoT: A secure block-based architecture for Internet of Things. IEEE Internet of Things Journal 6(2): 1961–1971.

[19] Kshetri, N. and Voas, J. 2018. Blockchain-enabled IoT. IT Professional 20(3): 17–24.

[20] Zeng, Y., Wu, H. and Zhang, Z. 2020. Secure and privacy-preserving data sharing in blockchain-based IoT systems: A survey. IEEE Communications Surveys & Tutorials 22(3): 2021–2044.

CHAPTER 12

Implementation of Blockchain for Secure Data Analysis in IOT Healthcare Systems

P. Venu Madhav,[1] *J. Sahithi*[2] and
P. Sa Dharmasastha Karthikeya[3],*

1. Introduction

The usage of communication tools in the healthcare system has brought unimaginable development in the problem areas related to detection and observation, which has potentially saved a number of people from life-threatening diseases such as cardiac failure, various types of cancer, and diabetes at their early stages, something which was previously deemed unachievable. Physicians have a critical role in the early diagnosis of Diabetes along with its management.

Focusing primarily on diabetic foot ulcer management, primarily requires vital knowledge of the major risk factors, frequent routine evaluation, along with preventive maintenance since if the condition is mismanaged or ignored it may result in serious consequences. Among several others, the most common factors leading to ulcer formation include uncontrolled diabetes, diabetic neuropathy (degeneration of the nerves of extremities), and gait imbalances due to foot deformity. The use of advanced communication systems can help us detect foot ulcers which were caused by diabetes and differentiate them from ulcers formed for a heavily weighted body. Many user trials with prototypes are being conducted and being developed to gain perspectives on

---
[1] Asst. Professor, Dept. of ECE, Prasad V. Potluri Siddhartha Institute of Technology.
[2] Asst. Professor, Dept. of Ophthalmology, Gayathri Medical College.
[3] Student, Department of BioMedical Engineering, Vignan University.
* Corresponding author: dharmasastha.v.biomed@gmail.com

diabetes management along its complications such as the foot ulcer based on the treatment requirements. The use of computer systems and various communication devices is a way to manage care along with detection and identification analysis of the data collected.

With the help of IoMT devices, which are a subset of IoT technology, medical equipment can now interact on their own across a network. IoMT networks are used to collect and send patient data to healthcare practitioners, often with little or no involvement from patients or their healthcare professionals.

Continuous infections are shown as chronic and require protracted therapy [1]. Patients with chronic illnesses typically stay at a clinic for a longer period of time for continuous monitoring. It's natural to have certain chronic potential diseases [2]. Diabetes already carries a high risk due to the frequent infections it frequently results in. In order for a patient with diabetes to have a normal life, they need be checked. When the body can't produce enough insulin or use it properly, the pancreas develops a permanent crack, which is what is known as diabetes [3, 4]. Numerous organs, including the eyes, nerves, and veins, can be harmed and rendered ineffective by high or low glucose levels; as a result, constant daily testing is crucial to prevent the loss of strength of the patient.

More constructions are anticipated to screen these people given the number of diabetic patients. The foundation of the diabetes monitoring system is regular blood glucose monitoring [7]. This suggests that both individuals and medical professionals can quickly check blood glucose levels at any time. Due to this, the 5G development, also referred to as the era of cutting-edge wearable associations, can achieve speedy communication, the expansion of major corporate boundaries, and hierarchical flexibility. Regardless, the focus of the current evaluation of this improvement is expanding the data provided [5, 6]. In this chapter, we develop a model that monitors and transmits data about diabetes patients to medical personnel and the patient's mobile device. The 5G cell network association is used by the mobile phone to transmit the sensor's data to the account station. In order to inform the patient, the information was obtained and further investigated.

2. Wearable IoT

They are also referred to as on-body IoMT devices. They keep track of a person's medical history and are connected to their body. Medical or consumer-grade wearable technology is available. Health information, such as a person's heart rate and blood pressure, are recorded by on-body IoMT. Consumer-grade smart gadgets can be used to track wellness and health metrics without a doctor's supervision.

3. Prototype and Working

The proposed prototype is of a bilayer flexible foot insole consisting of micro flexible temperature sensors on the top layer and pressure sensors on the bottom layer to sense the temperature and pressure changes in the foot.

Monofilament sensors are used to access the nerve signals of the individual if they are diagnosed with diabetic peripheral neuropathy. They contain of nylon filaments of varying thicknesses, when pressed against the foot the filament buckles a specific force indicating to perceive pressure, reduced sensation indicating nerve damage.

Continuous monitoring is the best available modality in healthcare as there is an almost negligible involvement of the individual to record the parameters of Pressure, Temperature, Nerve activity, Blood flow and the Glucose levels of the individual.

The temperature and pressure of the individual are to be measured through the sensors embedded in sole, the parameters of nerve activity, including blood flow are to be detected by the ankle and the glucose level parameter can be taken from the Continuous Glucose measurement (CGM) system where the sensor with a subcutaneous needle is placed near the hand or abdomen to manage their diabetes [8]. The entire data collected by various sensors can be transmitted to an application using Bluetooth which is of low energy consumption.

Predetermined threshold values for each of the taken parameters are set for the changes and when these parameters are reached above the pre-set threshold, an alert can be sent to individual or family members. The same data can be shared with health care professionals or their family physicians to remotely monitor the health of the individual or in the case of an emergency.

Figure 1 below represents a simple region where the occurrence of ulcer is first noticed in the case of diabetic foot.

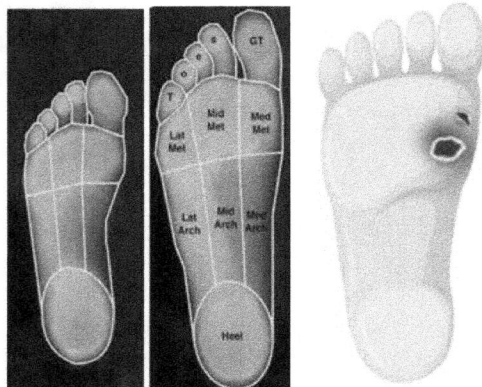

Figure 1. Depiction of a healthy foot and Ulcered foot.

One of the complications that accompanies diabetes is the presence of Diabetic Foot Ulcers. This is a chronic, non-healing, wound which affects the gait, posture, pressure and the temperature of the foot. It is caused due to a combination of factors like neuropathy (nerve damage), poor blood circulation, and diabetes.

The regions of the foot mentioned above get primarily affected for two reasons one is that it has a constant pressure of the weight of the patient having diabetes and the other is frequent disruptions in pressures of the foot on the ground where the skin comes in contact to the ground. With these circumstances, it is quite impossible to understand the major cause of wound occurrence.

Proper inspection and the management of diabetes on a regular basis is the key to preventing ulcer formation, unfortunately, the observation is subjective, and it depends on the individual to check for any visual changes on the foot.

HSI Hyper Spectral Imaging is actually a non-invasive and painless technique for differentiating pathological tissues from healthy ones. This is a very extensive process for the diagnosis or detection of a diabetic foot ulcer since the current approach through biopsies are conventional methods and along with being invasive, causes permanent wounds on the foot apart from other complications. The proposed device's integrated application receives the results from a metadata analysis [10] which measures several biomarkers by HSI (from all the sensors of the device) for estimating the health status of the individual by cross-comparing the values of the normal range which are already coded into the system. The goal of the analysis and cross-verification is for the identification and forecast of the disorders related to diabetes to some extent.

The sensors are ideally placed in six areas of the foot which are known to be prone to pressure, shear force and temperature changes. Clinically a 2.2°C of temperature rise over a point of the foot consistently over 48 hours has a great chance of developing into a diabetic ulcer, so as to prevent skin breakage, it is always advisable to offload the pressure on that area. The main limitation of thermal imaging is that is extremely sensitive to environmental conditions. In our case, we optimise an NIR (Near-Infrared) sensor to yield a discrimination analysis of the parameters of blood flow monitoring and tissue health detection of less than 100 mK differences and of spatial resolution of 10 pixels/com.

Thermal maps and pressure maps can be constructed using these data points for every hour and can be stored for further analysis. The prototype includes an accelerometer to measure the distance precisely before the pressure or temperature has reached its peak. This metric is used to determine the distance to offload the pressure on a regular basis. The processing stage provides a very accurate analysis of the parameters.

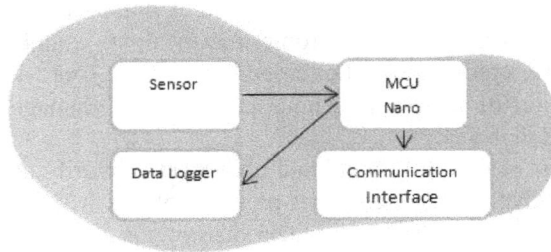

Figure 2. Layout of sensor placement.

The result of the analysis of the various parameters is combined into one result which will be presented in layman's terms and will be sent to the individual through a transmitter to an application.

Figure 3 is the schematic connectivity of the whole circuitry and components interfacing. The circuit diagram provided below also represents the pin connectivity of the overall modules.

Table 1 is the comparison of skin temperatures in the feet and the Ankle Brachial Index obtained in non-diabetic peripheral neuropathy (Non DPN, n = 46) and diabetic peripheral neuropathy (DPN, n = 39) patients.

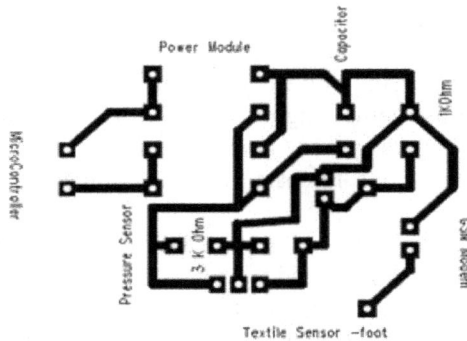

Figure 3. PCB Layout for modules interconnections.

The proposed model is for a remedial system for foot ulcers where we use embedded technology to project UV rays which act as the antibacterial projection device. A vibrator is used to oscillate the foot skin to revive it. The Peltier device is used to heat and cool the wound to further rejuvenate the foot skin to sustain a faster recovery. The HC-05 Bluetooth device transmits the instructions which are decoded by the Android application installed on the phone. These instructions are decoded in the form of audio-visuals and provided to the users for a better understanding and usability. Each session is created for multiple sessions of heating, and cooling using Peltiers. The vibration of the skin part is done using vibrators and the antibacterial part

Table 1. Comparison of foot temperature at various positions.

	Measures		Non DPN	DPN	Differences	95% CI	Effect size
ST (°C)	Hallux	R	32.33 ± 1.79	30.12 ± 2.26	2.21*	1.33–3.08	1.11
		L	32.56 ± 1.77	29.38 ± 2.55	3.18*	2.25–4.12	1.49
	Lesser toes	R	31.93 ± 1.97	29.43 ± 2.57	2.50*	1.52–3.48	1.12
		L	32.11 ± 1.78	28.92 ± 2.43	3.19*	2.28–4.10	1.54
	MMH	R	32.25 ± 1.51	30.53 ± 1.71	1.72*	1.03–2.42	1.08
		L	32.54 ± 1.54	30.08 ± 1.93	2.46*	1.71–3.21	1.44
	LMH	R	31.95 ± 1.57	30.33 ± 1.77	1.62*	0.90–2.34	0.98
		L	32.14 ± 1.62	29.73 ± 1.94	2.41*	1.64–3.18	1.38
	MidArch	R	32.50 ± 1.27	31.06 ± 1.41	1.44*	0.86–2.02	1.09
		L	32.75 ± 1.35	30.55 ± 1.63	2.20*	1.56–2.84	1.50
	MedArch	R	32.29 ± 1.38	30.65 ± 1.78	1.64*	0.96–2.32	1.05
		L	32.49 ± 1.52	30.02 ± 1.84	2.47*	1.74–3.19	1.49
	LatArch	R	32.21 ± 1.54	30.34 ± 1.85	1.87*	1.15–2.60	1.12
		L	32.43 ± 1.55	29.78 ± 1.96	2.65*	1.89–3.41	1.53

* Difference of Non-DPN and DPN (Real values)

by UV rays. The overall operation is synchronised using the microcontroller. Tailor-made programming can be done which makes it way more efficient and flexible for further changes and scalability. The overall process deals with multiple stages such as image pre-processing, segmentation, feature/texture extraction, and image classification. The pre-processing is used to enhance the quality of the input image by removing the noise.

4. Data Transfer in the IoMT Network

Healthcare businesses constantly struggle with security issues, whether it's protecting patient information or the IoMT infrastructure itself. The current healthcare systems also take advantage of IoMT where industrial sensors and actuators are used as wearable devices to collect users' physiological data, such as blood pressure, ECG, temperature and so on. In such a scenario, the data generated from industrial healthcare systems are often delivered or transmitted to a patient's local gateway or edge devices to perform data processing and aggregation, and then forwarded to the cloud for long-term storage, and furthermore also used by healthcare providers for real-time diagnosis and analysis. However, in the present healthcare ecosystem, the devices and sensors continuously monitor, communicate, and exchange information over an insecure public channel.

The entire healthcare system is also subject to a number of security vulnerabilities because of the constant connectivity of devices, including data manipulation, denial of service, eavesdropping, impersonation, man-in-the-middle, and replay attacks. Data tampering can result in inaccurate diagnoses, potentially putting the patients under surveillance in life-threatening situations. This raises serious issues in the healthcare sector. The security of an IoMT network is enhanced by software that automatically checks resources and identifies risks on the network. When a new device tries to connect to the department's IoMT network, this software can send an alarm. Such software works to prevent intruders and security threats from entering, damaging the network, and perhaps compromising the medical facility.

The systematic architecture of the proposed Blockchain framework is made up of 3 layers:

1) Industrial Healthcare system

2) Edge-Blockchain layer; and

3) Cloud-Blockchain layer.

1) **Industrial Healthcare system layer:** Important patient health data is continuously gathered at this layer using a variety of IoT-based healthcare tracking systems and implantable medical devices (such as temperature sensors, glucose monitors, and heart rate devices). These devices are referred to as lightweight nodes since they can only store and analyse a small percentage of the data on the blockchain due to their low resources and computational power (LNs).

2) **Edge-Blockchain Layer:** This layer is known as a full node and is made up of powerful nodes such as data analysis servers, industrial computers, edge computing servers, and so on (FN). The peer-to-peer network is constructed using edge devices geo-distributed amongst primary and urban health centres. Each patient is assigned an edge service node, which communicates with the cloud for long-term backup and storage and gathers, processes, and raises alerts in emergency scenarios.

3) **Cloud-Blockchain Layer:** The third tier is the cloud-blockchain layer, which consists of several cloud suppliers and data centres (DCs). These DCs are in charge of offering services to customers (such as computation, processing, and so on).

5. Integration of IoMT and Blockchain

Blockchain is the mechanism where the data is encrypted and restricts users from unauthorised access to encrypted data and is available with a key for authentic users to access the data for the authoritative users. This is one of the primary usages for the security of the patient's confidential health information which is one of the intricate parts of the Health care system throughout the world [10, 11]. When compared to the Blockchain, the IoMT systems have various parameters and a variety of architectural structures. The main challenge in integrating the blockchain with IoMT systems is that each sensor's data is collected and saved differently, making it difficult to modify and digitalise the data after it has been collected. This modified data will then need to be sent to the blockchain for encrypted analysis and representation as per a person's basic understanding through a BLE to an app.

There are basically three types of communication in this suggested framework: communication from medical devices to IPFS cluster nodes, communication from IPFS cluster nodes to smart contracts, and communication from smart contracts to the blockchain network.

Medical-device-to-IPFS Cluster Node Communication

In the paradigm, this communication is in charge of achieving two distinct goals. The registration of patients and their medical equipment is the primary goal. Prior to communication in the primary IoMT blockchain network, the second goal is to authenticate the medical devices.

IPFS Cluster Node-to-Smart Contract Communication

In order to ensure privacy in the IoMT blockchain network, this communication is in charge of synchronising the authentication and authorisation of the data from medical devices and their mapping.

Smart Contracts-to-Blockchain Network Communication

After successful authentication and permission, this communication is in charge of distributing the information into the blockchain network to enable secure data transmission between various agents (in this case patients and their doctors) in the IoMT blockchain network. The IoMT blockchain network's privacy is guaranteed by this communication.

6. Security and Privacy in IoMT via Blockchain

The security and confidentiality of the IoMT devices proceeds by the following tasks:

IoMT Wearable Device Verification

The IPFS cluster node closest to each IoMT device that supports smart contracts is used to register and authenticate the device as belonging to a particular patient in the IoMT blockchain network. The medical devices initially register with an IPFS cluster node that supports smart contracts. Once the identity of the medical equipment has been protected in an IPFS cluster node transaction, it is sent to the IoMT blockchain network to form a block. All additional peers in the IoMT blockchain healthcare network receive the blocks at this point for further access. Any devices attempting to communicate with the system must first authenticate and then send their security credentials to the IPFS cluster with smart contracts enabled.

Security of IoMT Devices

In the IoMT blockchain healthcare system, the suggested framework simplifies access control security for medical devices. According to this framework, only devices that are successfully authenticated and registered using IPFS cluster nodes that support smart contracts can join the IoMT network. The IoMT network does not accept connections from medical devices that are not listed on the IPFS cluster node since they are unable to authenticate them. As a result, the chance of malicious medical equipment interacting with the IoMT network is lessened.

7. Generation of Keys for IoMT Devices and Patients

The suggested methodology generates private and public keys for medical devices using the Elliptic Curve Digital Signature Algorithm (ECDSA) [10, 11]. The algorithm is chosen after comparing the several encryption techniques provided by The Rivest-Shamir-Adleman algorithm is more secure and requires fewer resources than the ECDSA technique (RSA). The Bitcoin framework has successfully employed the same algorithm [12].

Functionality of IoMT Devices

This subsection briefly explains the working of the proposed framework which is graphically illustrated in Fig. 4. The suggested model consists of two primary stages: the initialisation level, which is in charge of registering patients and their medical devices, and the second level, which is in charge of authenticating patients' medical devices. The patient initialisation level makes it easier for new patients to register in the IoMT healthcare system and ensures that each patient may be identified specifically. The next step is to register a patient's medical device with the IoMT healthcare system after they have enrolled as a new patient. During the device registration process, smart medical devices must register with the IoMT healthcare network. These medical devices are registered against the corresponding (previously registered) patients in order to join the IoMT network. The registration of the medical device is followed

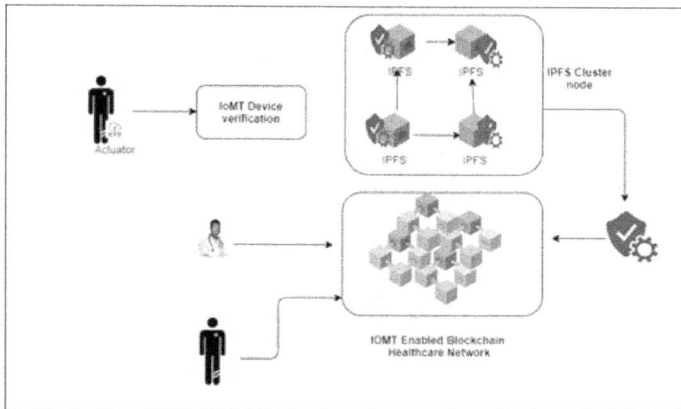

Figure 4. Proposed model of Blockchain IoMT-enabled Diabetes management.

by the patient's authentication, during which the IPFS cluster node with smart contracts enabled verifies the medical equipment. A certificate for the medical device's IPFS cluster node authentication is provided at the registration level. The medical equipment will be examined using their authenticity certificate at this stage. The medical gadget can join the IoMT healthcare system if all authentication requirements are met. This level ensures that the IoMT healthcare system can only be accessed by authorised medical devices.

8. Conclusion

The major goal of this chapter is to create a smart sole that will allow pressure readings from various foot pressure points to be systematically analysed and cross-checked against a database of predetermined parameter values. This may result in the analysis's findings being based on clinical information gathered through the smart sole. Multiple pressure sensors are incorporated into the sole as part of the design technique in order to track the pressure applied by both feet continually. The clinical data of parameters exerted on the foot through the mobile app is interfaced with the visualisation of prediction analysis and imported into the web app. To address the security and privacy of the patient data in an IoMT network, the framework is divided into two levels such as the Initialisation level and the Authentication level. The patient and their IoMT devices are registered at the initialisation level into the IPFS cluster node that supports smart contracts, and their patient identification and device identity are entered into the blockchain network. The patient ID, device ID, and device public address are mapped at the authentication level and shared with the IoMT blockchain network. Additionally, a distributed off-chain data storage layer that satisfies the requirements of secure storage management is provided by the IPFS cluster node. The patients' individual device information is submitted as a transaction to the blockchain network following successful registration and authentication in order to maintain anonymity in the IoMT network. Our future plan includes extending this work in the real-time application for the IoMT healthcare system by adding a larger number of peers and their IoMT devices.

References

[1] Global Diabetes, Diabetes.co.uk "Diabetes in India", https://www.diabetes.co.uk/globaldiabetes/diabetes-in-india.html, January 2019.

[2] The Times of India. Diabetic patients: Kerala tops the list of Indian states. https://timesofindia.indiatimes.com/city/kochi/diabetic-patients-kerala-tops-list-of-indianstates/articleshow/61974164.cms, Dec, 2017.

[3] Manohara, Pai M.M., Kolekar, S.V. and Pai, R.M. 2019. Development of smart sole based foot ulcer prediction system. 2nd International Conference on Intelligent Communication and Computational Techniques (ICCT), Sep, 2019.

[4] Patel, S., Patel, R. and Desai, D. 2017. Diabetic foot ulcer wound tissue detection and classification. International Conference on Innovations in information Embedded and Communication Systems (ICIIECS).

[5] Ostadabbas, S., Saeed, A. and Nourani, M. 2012. Sensor architectural tradeoff for diabetic foot ulcer monitoring. 34th Annual International Conference of the IEEE EMBS San Diego, California USA, September, 2012.

[6] Kumar, R. and Tripathi, R. 2021. Towards design and implementation of security and privacy framework for Internet of Medical Things (IoMT) by leveraging blockchain and IPFS technology. J. Supercomput. 77: 7916–7955. https://doi.org/10.1007/s11227-020-03570-x.

[7] Kotronis, C., IRoutis, I., Politi, E., Nikolaidou, M., Dimitrakopoulos, G., Anagnostopoulos, D., Amira, A., Bensaali, F. and Djelouat, H . 2019. Evaluating Internet of Medical Things (IOMT)-based systems from a human-centric perspective. Internet of Things 8: 100125.

[8] Mathavan,V., Venkatesan, R., Kumar, A.N.S. and Thulasidass, S. 2021. Health care devices for diabetic patient monitoring using IOT. pp. 1–6. 2021 International Conference on System, Computation, Automation and Networking (ICSCAN), Puducherry, India, doi: 10.1109/ICSCAN53069.2021.9526458.

[9] Kumar, R., Marchang, N. and Tripathi, R. 2020. Distributed off-chain storage of patient diagnostic reports in healthcare system using IPFS and blockchain. 2020 International Conference on Communication Systems & Networks (COMSNETS). IEEE.

[10] Kumar, R. et al. 2022. Permissioned blockchain and deep learning for secure and efficient data sharing in industrial healthcare systems. IEEE Transactions on Industrial Informatics 18.11: 8065–8073.

[11] Kumar, R. and Tripathi, R. 2020. A Secure and Distributed Framework for sharing COVID-19 patient Reports using Consortium Blockchain and IPFS. 2020 Sixth International Conference on Parallel, Distributed and Grid Computing (PDGC). IEEE.

[12] Kumar, R. and Tripathi, R. 2020. Secure healthcare framework using blockchain and public key cryptography. Blockchain Cybersecurity, Trust and Privacy (2020): 185–202.

Index

||||||||||||||||||||||||||||||||||||||||||||||||||||||||||||||||||||||||||||||||||||||||||||||||||||||||||||||||||||||||||||||||||||||||||||||||||

For Product Safety Concerns and Information please contact our EU representative GPSR@taylorandfrancis.com
Taylor & Francis Verlag GmbH, Kaufingerstraße 24, 80331 München, Germany